TIME FRAMES

ALSO BY NILES ELDREDGE

PHYLOGENETIC ANALYSIS AND PALEONTOLOGY
(with Joel Cracraft)

PHYLOGENETIC PATTERNS AND THE EVOLUTIONARY PROCESS
(with Joel Cracraft)

THE MONKEY BUSINESS

THE MYTHS OF HUMAN EVOLUTION
(with Ian Tattersall)

LIVING FOSSILS
(with Steven M. Stanley [eds.])

TIME FRAMES

THE EVOLUTION OF
PUNCTUATED EQUILIBRIA

NILES ELDREDGE

PRINCETON
UNIVERSITY
PRESS
PRINCETON
NEW JERSEY

Princeton Science Library paperback edition published by
Princeton University Press, 41 William Street,
Princeton, N.J., by arrangement with Simon & Schuster, Inc.

First Princeton Science Library edition printing
(with corrections), 1989

Text design by Edith Fowler

Cover design and photograph by Laury A. Egan

Library of Congress Cataloging-in-Publication Data
Eldredge, Niles.
Time frames : the evolution of punctuated equilibra / Niles Eldredge.
p. cm. — (Princeton science library)
Bibliography: p.
Includes index.
ISBN 0-691-02435-9 (alk. paper)
1. Evolution. I. Title. II. Title: Punctuated equilibria.
III. Series. QH371.E44 1989 575—dc20 89-34323

The author is grateful for permission to reprint material from the following sources:

Theodosius Dobzhansky, Genetics and the Origin of Species, *1937,*
copyright © 1982 by Columbia University Press. Used by permission.

Ernst Mayr, Systematics and the Origin of Species, *1942,*
copyright © 1982 by Columbia University Press. Used by permission.

Thomas J. M. Schopf, ed., Models in Paleobiology,
copyright © 1972 by Freeman, Cooper & Company. Used by permission.

Frederick J. Teggart, Theory and Processes of History, *1925, copyright © 1977*
by the Regents of the University of California. Used by permission.

Frontispiece: Middle Devonian rocks along the shore of Lake Erie in western New York.

Princeton University Press books are printed on acid-free paper, and meet the
guidelines for permanence and durability of the Committee on
Production Guidelines for Book Longevity of the Council of Library Resources

Manufactured by Princeton University Press in the USA

To Michelle, Rich, and Dave—
fellow voyagers through ancient seas

CONTENTS

LIST OF ILLUSTRATIONS

PREFACE

Charles Robert Darwin stands among the giants of Western thought because he convinced a majority of his peers that all of life shares a single, if complex, history. He taught us that we can understand life's history in purely naturalistic terms, without recourse to the supernatural or divine. Organic evolution remains one of the most profound notions of modern science, ranking with the basic laws of physics and with the general notion that the entire universe, certainly including our planet earth, has had an almost unimaginably long history.

Yet the media these days are full of announcements of Darwin's summary removal from the pantheon of Great Thinkers. Scientists, it is said, have been almost daily proving Darwin wrong. Compounding the problem has been the recent revival of "creationism," modern versions of which ("scientific creationism") echo the cry that Darwin was all wet—and challenge as well many other well-ensconced scientific tenets, such as the well-established age of the earth: 4.5 billion years.

Well, Darwin is by no means passé. Creationism is a wolf in sheep's clothing, biblical literalism simply (and clumsily) cloaked in the garb of science to evade constitutional injunctions that preclude religion in public-school curricula. And within science, Darwin's argument that life has had a history—has *evolved*—remains intact as a profoundly solid idea.

What *is* going on in evolutionary biology today is rather more subtle. Science is an ongoing collision between our ideas and our experience in the real, material universe. Though it sounds cynical, we cannot expect to make progress in our understanding of the nature of things unless we can show that what we thought we knew is in some sense wrong, or at least incomplete. It is the job of all scientists to ask new questions, to be sure, but also to keep a weather eye on old answers to see if they still hold up in the light of new observations. That evolution occurs no biologist worthy of the name doubts. But many biologists these days do openly wonder how complete and accurate our grasp of the mechanics of the evolutionary process really is.

Darwin, naturally, offered his thoughts on *how* evolution takes place. This is the special province of evolutionary theory. If it is true that an influx of doubt and uncertainty actually marks periods of healthy growth in a science, then evolutionary biology is flourishing today as it seldom has flourished in the past. For biologists collectively are less agreed upon the details of evolutionary mechanics than they were a scant decade ago. Superficially, it seems as if we know less about evolution than we did in 1959, the centennial year of Darwin's *On the Origin of Species*. Back then it seemed that Darwin was more nearly correct than he had often been given credit for in the decades since he announced his theory. Particularly, by 1959, the so-called "synthetic theory of evolution"—essentially Darwin's notion of evolution by natural selection smoothly integrated with a basic understanding of heredity developed only in the twentieth century in the post-Darwinian science of genetics—was accepted as *the* theory of evolution.

But it is not true that we know less about evolution than we did in 1959. We actually know a good deal more. In retrospect, some of Darwin's basic notions, integrated with genetics in the modern synthesis, seem a bit incomplete and inaccurate. For example, it should come as no real surprise that many of the recent discoveries in molecular biology have revealed a host of structures and functions within and between strands of DNA, the basic molecule of heredity whose structure became known only in the 1950s. Some molecular biologists, such as Gabriel Dover of Cambridge University, think they have evidence of processes operating within the genetic material itself, processes that affect the nature of genetic information passed down from generation to generation. Conventional evolutionary theory sees the modification of genetic information as the heart and soul of evolutionary change—but insists that Darwinian natural selection is by far the most powerful, if not the sole, agent mediating such change. Dover and his colleagues propose an additional set of mechanisms of change at a lower (molecular) level. Few if any evolutionary biologists claim that natural selection is unimportant; but some *do* argue that it is not the only important mechanism of evolutionary change.

Similarly, the theory of "punctuated equilibria," with its related notion of selection operating at levels higher than the organism (the level of Darwinian natural selection, which today is defined as "differential reproductive success" of organisms within a population), represents a rethinking of evolutionary processes based, ultimately, on the realization that a purely Darwinian, or "synthetic," view conflicts to some extent with our observations of nature. In this case, "nature" is not an array of molecules in some high-tech laboratory apparatus, but rather the pattern of change (and, critically, *lack* of change) we typically see when we look at fossils through a sequence of rock.

This book is my version of the story of "punctuated equilibria" —what it is, how it came to be, and what I think it means for modern evolutionary thinking. The development of the theory of punctuated equilibria is, I believe, utterly symptomatic of a general upsurge of critical analysis going on in virtually every corner of biology, including molecular and population genetics, developmental biology, systematics (the study of relationships and classification of all organisms), ecology and paleontology. In telling the story I have tried to capture the thought processes and the feelings that go into the formulation and execution of a fairly typical scientific study—as well as some of the more complex behavior that transpires once an idea begins to catch on, and once that idea requires further elaboration as its implications begin to emerge. It is a story, at the outset, of simply not finding the expected, and later, of the complex, if not totally tortuous, path that led on from there.

Punctuated equilibria still strikes me as an exceedingly simple idea: at base, it says that once a species evolves, it will usually not undergo great change as it continues its existence—contrary to prevailing expectation that indeed does go back to Darwin (and even beyond). Such species stability, at any rate, seems more the rule than the exception to judge from the fossil record. We concluded that most anatomical change in evolution seems to go hand in hand with the origin of new species—species being separate and integral reproductive communities of sexual organisms. There were, as I shall make clear, other ideas involved as well. We posed no new theory of speciation—that is, *how* a new species evolves. We found standard theory perfectly adequate in that regard, and in any case not a subject we could tackle from the standpoint of the fossil record. We did suggest that species could (and do) show the same sort of differential success in the game of life as organisms do. Simple and, I think, unobjectionable as these propositions are, they have been widely misinterpreted and, I would say, in some instances misrepresented. The furor that developed (in several quarters and at different times) tells us that something more is afoot than the straightforward objective evaluation of a set of postulates about the nature of things—the standard, if distorted, view of scientific behavior. Passions have been aroused, including our own, a phenomenon in itself informative to anyone who would understand how scientists do their thing. But the intensity of the reactions to punctuated equilibria, both pro and con, implies more than mere scrapping among territorial evolutionists: the debate has revealed some very deep disagreements over the very way we should be looking at the organic world.

From the very beginning, as we were working on the 1972 paper that coined the term and propelled the idea into general notoriety, Stephen Jay Gould and I have differed to some extent on the significance, the implications—and even, on occasion, some aspects of the basic content—of "punctuated equilibria." Though, of course, our positions are (and always have been) closely similar compared with opposing views, we have continued to debate fine points between ourselves and our close, like-minded colleagues—a necessity if we are to continue to grow. Neither Steve nor any other colleague mentioned here is to be expected to agree entirely with my version of the punctuated equilibria story. But I thank them all—like-minded and not—for a most stimulating past fifteen years.

1 EVOLUTION NOWADAYS

UTMOST CLARITY

THE CALIFORNIA coastline is volatile. The huge Pacific chunk of crustal plate surfaces as a thin rind of the North American continent, a sliver of rock running from Mexico up through L.A. and San Francisco before slipping once again below the waves. In the last 120 million years this outer rim of California coast has moved north perhaps as much as 350 miles—a rate of .2 inches per year. But that is only the average rate: what really happens is far more sporadic. Tension constantly mounts as the Pacific plate tries to slip north past the rest of the continent. Periodically the tension is released, and the sharp snaps of major earthquakes provide most of the movement along the famous San Andreas fault system. The net rate may be slow, but the movement itself comes in small steps that are often sudden and violent.

The rocks along that outer Californian rim are a mixed lot. Some are hundreds of millions of years old, accumulations of muds and sands that piled up when the crust was still submerged. There are balloon-shaped intrusions of granite—molten, igneous magma injected into the surrounding sediments. The magma has long since cooled to form massive topographic lumps, such as Bodega Head, 60 miles north of San Francisco. And there are younger sediments, some marine and others the accumulations of lakes and rivers, that have filled the basins and inlets which have appeared off and on as the earthquakes kept shifting the land around. Some of these younger sediments are jammed with fossils.

Not all the movement along the San Andreas is lateral: the rocks are sometimes shoved up or sunk down. The younger silts and sands have hardly had time to become firmly cemented into hard rock, and California is as notorious for its landslides as it is for its earthquakes. As a terrain, it is young and unstable—a difficult circumstance for a homeowner but a delight for the fossil collector.

The little town of Capitola, 10 miles south of Santa Cruz on Monterey Bay, has all the earmarks of a typical west-coast village. South of the beach, boardwalk and surfers, cliffs rapidly climb to an imposing height. Houses are precariously perched above (and even more dangerously tucked in *below*) the cliff wall and the beach is littered with huge chunks that keep spalling off the cliff face. Here is a vibrant, fresh exposure of the Purisima Formation—a thick sequence of sedimentary rock that surfaces periodically up and down the central California coast. Named for another town not too far from Capitola (geologic names are usually based on geographic features near a typical occurrence), the Purisima at Capitola is a light gray-green deposit of sands and silts that positively teems with fossils. Most of the beds are so soft that clamshells can easily be removed with a pocketknife.

"Purisima" is a truly apt name for these rocks, for they reveal the very essence of life's fossil record with utter clarity and simplicity. The fossils of the Purisima are mostly clams and snails: periwinkles, slipper shells, razor clams and so forth. A pale pearly pink, the shells have mostly lost the outermost, darkly colored layer living mollusks usually sport. They are a bit more delicate than the shells of recently dead organisms that litter strand lines along modern beaches. So a little care is in order in extracting a specimen from the friable, silty matrix. But otherwise the fossils are manifestly what they seem to be: simply the buried shells of clams and snails, plus crab carapaces, sand dollar skeletons, and the bones of whales, seals and porpoises. These animals—plus sea anemones, annelid worms and other creatures without hard skeletons to lend themselves to preservation—lived and died in some former coastal Californian sea. Occasionally their shells were buried in the bottom muds after death, in some cases trapped in rapidly formed deposits as the waves were agitated by a passing storm. Others simply accumulated quietly on the sea bottom, and were gently buried as the silts and muds built up.

It was something of a triumph for the human intellect to see fossils for what they are. It seems obvious to us today that an oyster jutting from a cliff, or a horn sticking out of a mountainside, *must* be

Cliffs exposing the Purisima Formation near Capitola, California.

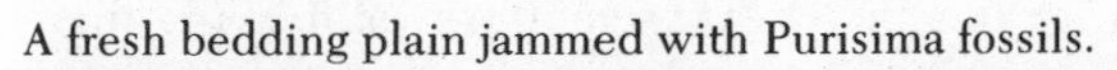

A fresh bedding plain jammed with Purisima fossils.

the preserved remains of some creature long dead. It was by no means so obvious to our remote forebears—nor is it intuitively obvious, probably, to a majority of the 4 billions of us on earth today. Herodotus, the Greek historian and traveler, thought that the presence of shells far inland in the hills encircling the Mediterranean implied the former presence of a much more extensive seaway. And Leonardo, with utterly characteristic verve and brilliance, realized that the fossil shark teeth he had picked up along his daily peregrinations must be simply what they seem: the teeth of sharks that had inhabited a now-vanished sea. But the received truth of the Middle Ages regarded fossils as either the work of the devil or the petrified remains of thunderbolts.

The Purisima Formation actually poses the problem in reverse. The shells are so fresh, so familiar, that it is not entirely obvious that they are *not* the shells of living creatures. Except, of course, that you stand with your back to Monterey Bay delicately prising specimen after specimen from a layer covered with 50 additional feet of sediment. Even more dramatic, perhaps, are the rocks and fossils of roughly the same age on the east coast. The western shore of Chesapeake Bay, along Maryland's Calvert Cliffs, produces a beautiful array of fossil shells—shells that periodically weather out, hit the beach and mingle with the shells of the mollusks still living there today. Sometimes only the realization that a particular species of clam or snail no longer lives in the Chesapeake tells you that the shell lying on the beach must be a fossil.

It is naturally a matter of simple curiosity to ask a few questions of this Purisima Formation. How old is it, for one thing? I have been calling rocks of Purisima ilk "young"—and they are, by reasonable geological standards. The Purisima at Capitola is Pliocene in age—which makes these rocks and fossils somewhere around 3 million years old. Most of the fossils I know best, in contrast, are from the Paleozoic, a grand division of geologic time that began close to 600 million years ago and ran for nearly 350 million years. Most of my fossils—trilobites—are about 400 million years old. That's why the Purisima seems relatively young to me. But, on the other hand, even someone steeped in the earlier phases of the 600-million-year-long history of complex life must admit that 3 million years is a respectable length of time. Three million years, yet the shells remain fresh—and distinctly modern in appearance.

And so we ask another question: how many of the species we

see in the Purisima—all the various crabs, mammals and mollusks —still live out there in Monterey Bay, or along the more southerly reaches of the California coast? The answer—simple as it is, and somehow as expectable as it may be, given the fresh look of the fossils—nonetheless is a bit shocking: perhaps as many as 90 percent. The vast majority of those 3-million-year-old species have modern names, names given to species of crabs, clams and snails of the modern California shallow-water marine invertebrate fauna. This is not to say that those Purisima species still considered alive today are identical in every feature to their fossil counterparts. But it does mean that the paleontologists who first found, studied and classified those Purisima fossils realized that the fossils were so similar to modern species that separate names for the fossils were simply not warranted. There just is no rationale, no purpose to be served in giving different names to such virtually identical creatures just because they are separated by 3 million years of time. Yet that *is* the natural propensity of paleontologists: collections of otherwise similar, if not completely identical, fossils tend to get different names for no reason other than their supposedly significant age differences. Evidently the similarities between Purisima fossils and living shells is *so* compelling that normal paleontological procedures just could not be followed.

And here is the profound significance of this modest exposure of Pliocene rocks and fossils at Capitola, California: the simple facts of the matter fly squarely in the face of the most deeply entrenched canons of the paleontological profession—for if evolution is anything, it is change through time. We have become convinced of the sheer *inevitability* of change given the mere passage of time. So deep are the roots of this belief that paleontologists have been searching with great frustration and largely in vain for "insensibly graded series"—perfectly, progressively intergrading series of fossils that exhibit gradual, directional change up through a sequence of layered rocks. "Insensibly graded series" is Darwin's phrase, but the expectation that change is inevitable given the mere passage of time goes back well before Darwin. Paleontologists have literally been caught between a rock and a hard place: Evolutionary theory, or so they have thought, predicted that most evolutionary change would be slow, steady, gradual and progressive. We would reasonably expect that, at least in exceptionally fossiliferous sequences, we should find sufficient examples to bear out this prediction. But paleontologists (and, as we shall see, even Darwin himself) realized that examples of gradual, progressive change are rare in the fossil record. Thus the conflict: the actual history of evolutionary events, though universally conceded to be but imperfectly recorded in the rocks, nonetheless has always seemed strongly at odds with the conventional predictions Darwin and nearly all later evolutionists (paleontologists and other biologists alike) maintained *must* be true.

Matching observations with predictions derived from abstract theory is, of course, the lifeblood, the central process of all science. Science is a system of ideas, a body of statements about the material universe. Only logical consistency and especially the evidence of our senses can serve as a check, a way of assessing the accuracy of our ideas about the world around us. It is a procedural norm of science—though, curiously enough, a dictum honored most often in the breach—that when one's thoughts about the way things are *supposed* to be conflict with what seems to be the actual truth of the matter, we reexamine those notions which led us to erroneous predictions. So we must ask why the conflict between expectations and simple phenomena such as the Purisima fossils has persisted so long. Part of the answer lies in the very problems Darwin had to face just to convince the rest of the world that life had in fact evolved.

DARWIN'S INHERITANCE

Darwin didn't invent the idea of evolution. Alfred Russel Wallace, who later independently came up with the idea of natural selection, apparently set out for the Amazon forests in the 1840s with both his companion Henry Walter Bates and a firm conviction that life had evolved. As Frederick J. Teggart pointed out in his *Theory of History* (1925), the very idea of evolution stems from the Cartesian vision of a universe in constant motion. According to Teggart (p. 133), it was an achievement of eighteenth-century science in general to see the modern world as the result of changes that took place in the past. Darwin inherited a legacy of general notions of historical change—and, as is well known, his own grandfather was among his direct forerunners in shaping a definitive picture of the evolution of life.

Teggart's analysis of evolutionary thought up to his day makes absorbing reading. As a historian, Teggart saw an analogy between the biological theory of evolution and similar notions in the social sciences. Diffidently disavowing any biological expertise, Teggart nonetheless delivered a stunning exegesis of biological theory—and a critique that would have been devastating had it ever been read by biologists. Charged with the task of developing a coherent social-studies curriculum for secondary schools in the early decades of this century, Teggart saw this social science in disarray. There seemed to be no coherent theory that united views of social organization with notions of how social systems change through time. And in particular, Teggart saw the two dominant approaches to the study of historical change—the work of historians on the one hand and that of evolutionarily minded sociologists and anthropologists on the other—as both severely flawed. The problem, simply, was the failure to integrate the actual *events* of history with general theories of historical change.

Teggart saw the description of the events of history as the special province of historiographers, people actually engaged in writing what we conventionally call "history." Historiography done well makes compelling reading, but Teggart was concerned with the lack of what we might call "rigor" in the output of the typical historian. There was then (and still is only emerging now) little in the way of comparative history, in which patterns of change among two or more societies might be contrasted and their similarities and differences recorded with an eye to developing general theories of social change through time. Rather (or so it seemed to Teggart), the goal of history seemed far more the glorification of a nation's pride, and historiography too often read like literary criticism, with too much emphasis on attempting to divine the motives underlying the actions of singular historical individuals.

But "evolutionism" seemed hardly any better. Teggart saw the general idea of evolution as linked with the notion of progress, and indeed it has recently become all the rage to point to Darwin's near infatuation with the view that evolution typically represents progressive change (see, for example, Jeremy Rifkin's generally compelling argument along these lines in his *Algeny,* and a similar, if more restrained, discussion in Eldredge and Tattersall's *The Myths of Human Evolution*). But the connection between evolution and progress was made long before Darwin. For example, Teggart writes (p. 133):

> The theory of "evolution" rests, in the first place, upon the assumption that "progressive change" is "natural" and to be taken for granted, and that the aim of this progressive movement is the attainment of perfection. In the judgment of Erasmus Darwin [Darwin's grandfather] (1794), "it would appear that all nature exists in a state of perpetual improvement by laws impressed on the atoms of matter by the great Cause of Causes; and that the world may still be in its infancy, and continue to improve forever and ever."

But as a further and perhaps more fundamental objection to eighteenth- and nineteenth-century notions of evolution, Teggart worried about biologists in effect taking *time* for granted. "The assumption [Teggart writes, p. 134] that change is invariably slow, gradual, and continuous entails the very important condition that we may neglect the element of time." He cites the great French biologist Jean-Baptiste de Lamarck precisely to this effect. In his 1802 *Hydrogéologie,* Lamarck wrote: "For nature, time is nothing. It is never a difficulty, she always has it at her disposal; and it is for her the means by which she has accomplished the greatest as well as the least of her results." But there is a danger here. The peril that arises when one ignores the actual events of history derives from this simple assumption that given enough time, all things may eventually happen:

> When, however, it is assumed that Nature always has unlimited time at her disposal, and that change is invariably slow and gradual, the statement is equivalent to the assertion that, in the study of evolution, the possibility of "events" may be ruled out of consideration. The dictum that "Nature never makes leaps" thus comes to be accepted as assurance that there never have been "events" in the history of the forms of life [Teggart, 1925, p. 135].

And somehow, to Teggart (and to many of us), the events of history, whether biological or social, must surely have something to tell us about how that change actually occurs.

Teggart was clearly on to something. Historiographers look at historical events but fail to generalize on them, to see the classes of similar events that might emerge to become patterns in need of general theories of social change. Historiographers didn't impress Teggart as particularly "scientific." On the other hand, social evolutionists had a theory which they uniformly were not checking against such classes, or patterns, of events, even though theirs was indeed a comparative approach. Steadfastly maintaining a general theory at odds with the facts of the matter—the actual events of history—is, if anything, even less rigorously "scientific" an approach to history. Yet the social evolutionists were not alone; according to Teggart, Darwin was doing it too:

> Darwin took over and urged insistently the principle that Nature never makes leaps—*natura non facit saltum*. It is to be noted, however, that he appears to have accepted the principle of continuity only in its genealogical form; he did not adopt the view commonly held by his predecessors that the series of existent [Teggart here means "modern"] life forms was also continuous. He regarded the historical series as alone representing the "natural order."* Hence, in his conception, Nature, in the course of time, moves only by slow, gradual steps, by slight, successive transitions. He was thus led to maintain that the number of intermediate forms which formerly existed *must* have been "interminable," "enormous," "inconceivably great."
>
> The crucial element in the presuppositions accepted by Darwin may be given in his statement that, as Nature can act only by short and slow steps, she can produce no great or sudden modifications.* Now the point to be observed, in relation to the present discussion, is that Darwin regarded this dictum as possessing a higher validity, for evolutionary study, than the facts of biological history. Indeed, he devoted a chapter of the *Origin of Species*—"On the Imperfection of the Geological Record"—to the argument that, since the available information in regard to the past is imperfect and incomplete, it may be set aside altogether in favor of the canon *natura non facit saltum* [Teggart, 1925, pp. 137–38. Asterisks (*) mark his footnoted citations to quotes from the sixth edition of Darwin's *Origin*.]

So, Teggart sees Darwin almost as a victim of prior intellectual commitment to the twin notions that to change is to progress, and that change is inevitable given the mere passage of considerable lengths of time. And it is true that Darwin devoted *two* chapters (9 and 10 of the first edition) to comparing his views on evolution with what was then known of the general nature of the fossil record. In an oft-quoted passage, Darwin wrote (*Origin*, first ed., p. 341):

I have attempted to show that the geological record is extremely imperfect; that only a small portion of the globe has been geologically explored with care; that only certain classes of organic beings have been largely preserved in a fossil state; that the number both of specimens and of species, preserved in our museums, is absolutely as nothing compared with the incalculable number of generations which must have passed away even during a certain formation; that, owing to subsidence being necessary for the accumulation of fossiliferous deposits thick enough to resist future degradation, enormous intervals of time have elapsed between the successive formations; that there has probably been more extinction during the periods of subsidence, and more variation during the periods of elevation, and during the latter the record will have been least perfectly kept; that each single formation has not been continuously deposited; that the duration of each formation is, perhaps, short compared with the average duration of specific forms; that migration has played an important part in the first appearance of new forms in any one area and formation; that widely ranging species are those which have varied most, and have oftenest given rise to new species; and that varieties have at first often been local. All these causes taken conjointly, must have tended to make the geological record extremely imperfect, and will to a large extent explain why we do not find interminable varieties, connecting together all the extinct and existing forms of life by the finest graduated steps.

He who rejects these views on the nature of the geological record will rightly reject my whole theory . . .

Here Darwin clearly links his theory with the expectation that the fossil record of life's evolution should be crammed with examples of intermediate forms. Earlier on, he remarked (p. 280):

But just in proportion as this process of extermination has acted on an enormous scale, so must the number of intermediate varieties, which have formerly existed on the earth, be truly enormous. Why then is not every geological formation and every stratum full of such intermediate links? Geology assuredly does not reveal any such finely graduated organic chain; and this, perhaps, is the most obvious and gravest objection which can be urged against my theory. The explanation lies, as I believe, in the extreme imperfection of the fossil record.

Thomas Henry Huxley, Darwin's "bulldog" (who is said to have remarked, upon reading the *Origin* for the first time, "How extremely stupid not to have thought of that"), enjoined Darwin to tread lightly with his insistence that evolution must always be gradual and progressive. "You have loaded yourself with an unnecessary difficulty in adopting 'Natura non facit saltum' so unreservedly" (letter to Darwin, 1859).

Some of Darwin's more recent defenders (among them Ruse, 1980; see also David Penny's article in the March 1983 number of the technical journal *Systematic Zoology,* and a commentary by F. H. Rhodes in *Nature,* 1983) have pointed out that Darwin included many passages in the *Origin* (especially in later editions) in which he did indeed acknowledge that evolution proceeds at a variety of rates. He even conceded that once they have evolved, the geological record seems to show a typical pattern of relative stability for many species. It is thus simply setting up a straw man, these critics claim, to label Darwin as a simpleminded "phyletic gradualist"—one who insists that evolution takes place by the slow transformation of an entire species through geological time. And it is certainly true that Darwin was a clever, knowledgeable man who was well aware of nature's multifarious complexities. It is also true that ensuing editions (particularly the sixth—the last and most commonly reprinted version available today) included a large number of additions as Darwin attempted to meet his critics' objections. Most famous, perhaps, was Darwin's grudging admission of the "inheritance of acquired characters" during an organism's lifetime alongside his beloved "natural selection" as a moving force in evolution. But for every quote one can dig out of the various editions that seems to support a more pluralistic, less rigidly "evolution is purely the slow, steady transformation of entire lineages" position, there are scores which plainly show that very view indeed to be the gist of Darwin's thought. Darwin was less a pluralist on these matters than some these days would have him be.

Yet Teggart once again points out the truly interesting lesson of Darwin's confrontation with the fossil record. Darwin's early scientific experience was primarily as a geologist, and much of what he had to say about the nature of the fossil record (summarized in the passage quoted above) was an accurate and insightful early contribution to our understanding of the vagaries of deposition and the preservation of fossils. But his Chapter 9 (first edition) on the imperfections of the geological record is one long *ad hoc*, special-pleading argument designed to rationalize, to flat-out explain away, the differences between what he saw as logical predictions derived from his theory and the facts of the fossil record.

Yet, as Teggart reminds us, Darwin actually admitted that the *known facts* of the geological record spoke out against his major prediction: that evolution on the whole must be slow, steady, gradual and progressive. To understand why Darwin was so adamant, and to shed some light on the persistence of this conflict between what we've been expecting to see and what we have actually been seeing all along, we must consider Mr. Darwin's theory in a bit more detail.

DARWIN'S TASK

Darwin came back from his five-year tour of duty on the H.M.S. *Beagle* in 1836. And he came back on the brink of conviction of the *fact* of evolution—that all creatures past and present are linked up in one grand, intricate genealogical array. It was "descent with modification" (Darwin actually never used the word "evolution" in his first edition of the *Origin*). The concept was simplicity itself: all life has descended from a single common ancestor, and along the way of this interminable process of ancestry and descent, creatures become modified. As lines of descent split, life has diversified; descendants within these disparate lines have inherited ancestral characteristics, duly passing them along in sometimes still-further-modified form to still later descendants. It was an idea, Darwin saw, that so clearly explained why there is a complexly layered pattern of similarity that unites all organisms. Dogs, wolves and jackals share a goodly array of anatomical and behavioral features. But they also share with bears and cats a more general similarity that seems to link them all into a larger group—a group mammalogists call the Order Carnivora. And these carnivores—these bears, dogs, cats, weasels, civets, hyenas, racoons and seals—are clearly all mammals. They have hair, mammary glands, four-chambered hearts, three little bones in the middle ear and placental development in the uterus. Darwin saw such ever-widening circles of group membership as clearly implying the natural origination of new features—items such as hair and mammary glands—at some specific point in the genealogical history of a group. Such successful innovations, he imagined, were simply handed down to all later descendants. Sometime after the invention of hair and the other basic, early features that all mammals retain today, a single lineage specializing on the consumption of other animals developed a highly specific shearing dentition—a successful innovation that has been passed on to many descendants (and lost by a few, such as the bamboo-eating giant panda). The trait helps define the mammalian Order Carnivora. Indeed, the only competing explanation for the order we all see in the biological world, this pattern of nested similarity that links up absolutely all known forms of life, is the notion of Special Creation: that a supernatural Creator, using a sort of blueprint, simply fashioned life with its intricate skein of resemblances passing through it.

And, of course, it was precisely this notion of divine Creation that furnished the explanation for all life—its very existence, its exuberant diversity and its apparent order—in Darwin's day. A naturalistic, materialistic account, one that saw present life as a product of a long history of natural rather than divine processes, was a truly radical and rather heavy notion. By the end of the 1830s, Darwin had already come up with a personally satisfying theory of *how* evolution occurs. And yet he waited, delaying publication, gathering up a vast encyclopedia of arguments and examples from nature to bolster his thesis. He was to wait nearly twenty years before announcing publicly his theory of evolution—his devastatingly strong argument that life must have evolved, and his cogent theory, natural selection, which supplied the evolutionary mechanism. And even then it was only because Alfred Russel Wallace nearly scooped him that Darwin sprang to the fore, overcoming his reluctance out of sheer fear of losing credit for his life's work.

Of all the putative reasons for Darwin's delay, surely the main one was the certain outrage such a materialistic view of the organic world would provoke. Part of Darwin himself was shocked by the idea, though he eloquently maintained in the closing paragraph of his *Origin* that "There is grandeur in this view of life, with its several powers, having been originally breathed into a few forms or into one; and that, whilst this planet has gone cycling on according to the fixed law of gravity, from so simple a beginning endless forms most beautiful and most wonderful have been, and are being, evolved." Whatever the reason for his procrastination, it is clear that Darwin felt compelled to make as airtight a case as he could muster before publishing his theory. That he succeeded admirably is beyond dispute: today's creationists notwithstanding, it is the received view of modern, Western-world science that life has evolved. The first printing of the *Origin*—some two thousand copies—sold out the very first day it appeared at the booksellers'. It was clearly an idea whose time had come.

How are successful new ideas—in science or, for that matter, in any sector of human thought—introduced? What was the magic ingredient Darwin had that his evolutionarily minded forerunners lacked? The quick and dirty answer is simply that Darwin supplied a cogent, plausible mechanism which explained *how* life can evolve. We have recently seen a supposedly parallel case in the earth sciences, in which the vision of the earth's crust as fragmented into some nineteen major plates constantly moving around the globe finally gives us both an integrated theory of the earth's internal "physiology" and a process that explains the earth's history. "Plate tectonics," the modern version of "continental drift," truly is the earthly equivalent of organic evolution. It is commonly said (see, for example, Anthony Hallam's informative *A Revolution in the Earth Sciences*, 1973, for a discussion) that a notion of a mobile earth could not be widely accepted until geophysicists came up with a believable mechanism that would allow continental crust to plow through the denser crustal materials of the oceanic basins. The older evidence for continental drift had been the matching up of continental outlines and fossil faunas and floras—patterns that spoke eloquently to a few brave souls (such as the Englishman Arthur Holmes and the South African Alexander du Toit). Such data fell on deaf geophysical ears.

When plate tectonics was invented in the early 1960s it took the geological profession by storm. In only a few years the idea became the guiding research paradigm of the entire field. The reason: some geophysicists were virtually forced to take a mobilist view of the earth if they were to make sense of some of their newer data. Remnant magnetism—weak traces of the earth's polar orientation in remote times—could be measured with newly devised, sensitive instruments. Some odd results began to accumulate: 400 million years ago, for example, the north magnetic pole, as seen from Great Britain, was somewhere out in the western Pacific. Either the pole had been wandering, or Britain had. When the magnetism of rocks of the same age was measured in North America, the pole position was likewise aberrant, but it wasn't where the British pole was! Only when you mentally push Britain and North America together and rotate them a bit do the poles coincide and go back where the single magnetic pole belongs—up around the geographic pole, the northern epicenter of the earth's rotational axis.

If magnetic poles don't really wander, they do reverse themselves: periodically the earth's magnetic field drops to zero for some as yet unknown reason. When the charge starts to build up once again, it sometimes comes in with reverse polarity: the earth is like a giant, bipolar bar magnet. Sometimes the north pole is positively charged, and the south magnetic pole is correspondingly negative. At other times, the north pole is negative, the south positive. Dragging magnetometers over the ridge that runs smack down the middle of the North Atlantic, a zebra-stripe pattern of + and − bands, parallel to the ridge and absolutely symmetrical on its two sides, surprised another set of geophysicists. When it was learned that the sediments on and just adjacent to the ridge are very young, and the sediments get older as you travel directly away from the ridge, the game was up: new crustal material must be forming at the ridge, cooling and sinking as it moves laterally away in front of ever more upwelling basaltic magma. While the basalt is cooling, the iron particles in the molten basalt take on whatever magnetic orientation the earth has at the moment. Times change, the poles flip and the next batch of crust has the reverse orientation.

Voilà. Geophysicists finally had the key to the puzzle. But had they, really? Theirs was the evidence that finally convinced *them* that huge segments of the earth's crust are moving around in relation to each other. But the *mechanism?* We still really don't have it. Compelling as it is, the magnetic evidence is really part and parcel of the very same kind of *pattern* geologists and paleontologists had been seeing for years: coal in Antarctica; the same species of 380-million-year-old trilobite in Ohio and the Spanish Sahara.

And the same holds for Darwin and *On the Origin of Species.* We often hear that Darwin's proposed mechanism—*natural selection*—was the missing gem, the key to his success where others before him had failed to convince their peers that life had evolved. Yet as rapidly as evolution itself was taken to the Victorian scientific bosom, many of Darwin's contemporaries had trouble with his proposed mechanism of natural selection. Many of the squabbles in evolutionary theory today still center around the strength and nature —the relative *importance*—of natural selection in the evolutionary process (see Michael Ruse's *Darwinism Defended* or my *Monkey Business* for a rundown on current controversies over natural selection). Selection was important to Darwin's argument, of course. But as ardent a selectionist as he was, as we have already seen, even Darwin retreated a bit to the point of letting the inheritance of acquired characters creep into his narrow range of candidates for evolutionary forces. It was important that he present some plausible mechanism of the evolutionary process, but just as plate tectonics ultimately caught on by the sheer weight of the evidence—literal patterns that could brook no other explanation save the supernatural —so was it Darwin's impressive marshaling of the facts of the organic world that really carried the day.

Nothing in biology makes sense except in the light of evolution, as the geneticist Theodosius Dobzhansky once wisely remarked. Darwin simply argued that the hierarchy of anatomical similarity, the startling resemblances otherwise dissimilar organisms share in their early embryonic stages, the tendency for similar, obviously related species to replace one another in adjacent geographic regions, plus the general progression of life forms through geological time virtually forced one to accept his notion of descent with modification. Coupled with natural selection—nature's way of achieving the sorts of results achieved by animal breeders—Darwin's argument was an irresistible one-two punch.

SPECIES FIXITY

But to really understand why Darwin was so insistent that evolutionary change is inevitable, a virtual necessity given the mere passage of time, we need only look at the biological perspective he was fighting against. "There are as many species as originally created by the Infinite Being," or so said the Swedish naturalist Linnaeus (Carl von Linné) in the eighteenth century. Linnaeus founded our modern system of hierarchical classification, based on the realization that there *is* order in nature. That there is a skein of progressive similarity linking up all organisms was appreciated by Aristotle, but the notion did not emerge as an important organizing principle in biology until Linnaeus and a few contemporaries started classifying animals and plants in earnest. And though a few of these eighteenth-century systematists had vaguely evolutionary notions, nearly all were devoutly and orthodoxly religious. They saw the order in their material, the grand pattern of similarity running through the entire organic realm, as evidence of God's plan of Creation. Thus the dictum: All species were separately created. And all species we see today were created in their present forms back in the Beginning—which was not especially startling, as the Genesis version of the event, those fateful six days, was reckoned as falling only 6,000 years before. Six thousand years is not an unimaginably long time to expect a species to remain stable.

Thus, in a very real sense, the very antithesis of "evolution" is the belief that species were separately created and have remained static, fixed for all the ages. It was "species fixity" that posed the strict alternative to "descent with modification." To establish the very idea of evolution, Darwin naturally gravitated to the other extreme in the spectrum of possibilities. As Ernst Mayr, one of the founders of the modern synthetic theory of evolution, pointed out in his *Systematics and the Origin of Species* (1942), Darwin never really did discuss the origin of species in his *On the Origin of Species.* Species are defined in many ways, but the concept that dominates modern thinking sees species as reproductively coherent communities, seldom if ever capable of interbreeding with other species (see Ch. 4 for much more on the nature of species and their origins).

But the opposite of a fixed species, the creationist view of species, is simply a lineage in flux. Evolution cannot happen if species are "fixed"; thus, since evolution seems to have occurred, species simply *cannot be fixed.* They must instead be labile. Darwin came to see species simply as stages in a stream of continual anatomical development. Yes, he supposed, it is true that the species of birds in the backyards of English homes all seem rather discretely different from one another. But, he ventured, at least the closely related ones surely were less different from each other in the not-too-distant geological past. And it seemed a safe bet that they would continue to become modified, to diversify, still more in the not-too-remote geological future. For why, evolutionists are occasionally wont to ask, should we expect evolution to stop just because we have happened upon the scene?

Darwin's view of species persists in some contexts right up to the present day. Species are often regarded as ephemera, of no particular evolutionary significance whatsoever. In contrast, the paleontological theory of "punctuated equilibria" affords one of the strongest of recent arguments that species are crucial flesh-and-blood actors in the evolutionary drama. Species are real entities, not just a passing stage in a continuous evolutionary stream. In a way, then, we have borrowed a page from the older, creation-minded naturalists: though species are *not* fixed in any rigid sense, the fossil record tells us that species tend to remain stable, thus readily recognizable, for truly formidable periods of time, often millions of years—far more than the 6,000 years allotted by Judeo-Christian cosmology.

NATURAL SELECTION

Animal husbandry goes way back—at least as far as Neolithic times, when the bones of domesticated animals first show up in archaeological digs in the Middle East. Somewhere along the way breeders started literally selecting out only those individual animals with the more desirable traits—higher milk production, a knack for laying more and larger eggs—and let only the favored few produce the next generation's cows, chickens, dogs and donkeys. Darwin, as is well known, spent a great deal of time studying the ins and outs of pigeon breeding, and used his experience in the very first chapter of the *Origin*. For Darwin simply saw nature as equipped with an automatic sorting system very much along the lines he himself had learned from his fellow pigeon fanciers.

Darwin's one great "eureka" came, he tells us, when he was riding along in his coach in 1838. Suddenly everything clicked and the idea of natural selection hit him like a thunderbolt. He had read the economist Thomas Malthus, whose 1798 pamphlet *On Population* laid out a few simple truths about birth and death. Malthus showed that population size will naturally increase geometrically, so that some factor must normally be holding natural populations in check. Otherwise any species (Malthus actually directed most of his analysis to human populations) would quickly overrun the earth. Ordinarily, the controlling agent would be the energy sources—food

—that all organisms need to keep going. Darwin's own example in the *Origin* was elephants: it would take a mere 500 years for a single pair of elephants to turn into a thundering herd 15 million strong—and that is allowing a reproductive rate of but one offspring per couple every 10 years. Alfred Russel Wallace, who came up with the very same idea while combating malaria in the Spice Islands, had observed a population explosion of mice in South America years earlier—an event that he integrated with Malthus to yield "natural selection."

Both Darwin and Wallace saw that there must be competition for limited resource supplies. And they also realized that with the exception of some pairs of twins, no two organisms within a population are ever exactly alike. Such natural variation ensures that the competition will be unequal: it will be in the nature of things for some organisms to be more vigorous, more viable, more "fit"—generally more able by dint of overall better health, or equipped with a superior version of some feature vital to survival—keener eyesight, swifter legs, sharper teeth. And that's all there is to natural selection: in the competition for resources, the variability in the population means that some organisms will be better equipped to survive—and thus to reproduce, to leave offspring to the next generation. On the average, the more fit will tend to leave more offspring. And though Darwin and Wallace of course shared the general biological ignorance of the mechanisms of heredity in their day, they both knew that organisms tend to resemble their parents. Thus the favorable traits their competitively superior parents had, the very traits that had helped get their offspring there in the first place, would tend to be passed along. The succeeding generations will tend to have those same traits; thus the superior traits are handed down and spread within the population. The stock is gradually changed as nature herself steps in and dictates which traits are to be favored for greater representation in the next generation.

Selection seemed to Darwin an inevitability—the ineluctable consequence of competition among unequals for limited supplies. Though it *sounds* very much like laissez-faire Victorian economics, Darwin and Wallace were in fact describing a simple dynamic of nature. Populations really *are* resource-limited. And there always is differential reproductive success within populations. Experiments, field observations and the inductive tools of mathematics have all been brought to bear on this simple process of natural selection. And all confirm its existence and powerful role in the economy of nature. Natural selection is no shopworn Victorian fantasy that better describes the affairs of businessmen than the workings of nature, as Jeremy Rifkin, for one, concludes in his *Algeny*.

Yet it is easy to see how simple is the leap from "natural selection is always there" to "change is inevitable." Darwin had an extremely powerful mechanism at his disposal, one that fitted in perfectly with the already rampant feeling that the universe is in constant motion and change is in the very nature of all things. Darwin saw natural selection as the moving force underlying the *modification* side of his "descent with modification." He saw organisms adapted to their environments, the better to make a living and to compete for those limited resources. Adaptation—design improvement through natural selection—was the central facet of Darwin's entire evolutionary vision. Adaptation was Darwin's triumphant answer to the ultimate Creationist challenge: how else must we explain the obvious *design* in nature should we not concede it to be the conscious work of a master Designer? Darwin's—and Wallace's—answer was simple: natural selection, inherent in the very scheme of things, is an antichance mechanism that simply orders natural variation. It is a purely natural, materialistic designer.

Thus selection will tend to modify adaptations when circumstances that had favored older forms of life change. And even if environments remain constant, there is always the possibility of improvement, strengthening existing adaptations rather than forging new ones. And here is the clincher: given the sheer eventual inevitability of environmental change, anatomical and behavioral change surely must follow close behind.

Thus a neat concatenation of factors forced Darwin to conclude that evolutionary change—rather than stability—is in the very nature of things. It was already in the air; the Victorian predilection for seeing change as "progress" certainly encouraged such a view. Darwin saw the very necessity of change as the only solid argument against "species fixity"—which in turn was the older and distinctly antievolutionary view of the earlier biologists. And he and Wallace described a natural mechanism which seemed to each—and to many of their successors—as virtually ensuring *some* measure of change as the generations rolled. Darwin himself may have been something of a pluralist, open to exceptions in his generalizations about Mother Nature, but thousands of biologists and paleontologists heeded his main lesson: evolutionary change simply must show up as we contemplate particular lineages of fossils. If we chase fossil oysters up a cliff, we should see them slowly change before our eyes. Sometimes we do. More often, as at Capitola in the Purisima Formation, we do not.

2 ANCIENT SEAS AND EVOLUTIONARY FANTASIES

THESE ARE unusual times. The continents stand high and dry, and though they occupy a mere 29 percent of the earth's surface —the rest given over to the seas—yet it is not the normal course of things for so much land to be exposed. It is far more usual to find the lands flooded, covered with a thin veneer of seawater. The great polar ice caps have locked up a tremendous volume of water, and sea level has been rising for the last 20,000 years or so as the latest glacial sheets retreated. But the ice caps, with their oscillatory growth which produced the continental glaciers of the Ice Age over the past 2 million years, have only relatively recently reached their present size. Over the last half-billion years the polar caps have generally been smaller, and there has simply been a lot more water to spread out over the earth's surface.

North America 380 million years ago would have been urecognizable to anyone familiar with the plains of Iowa and Kansas, the Rockies of the West or even the far more modest folded ridges of the southern Appalachians. Low, stubby hills did protrude along the more northern regions of the eastern continental margin, and the Adirondacks and northern Canada likewise stood above the surface. Much of the rest of what is now the United States and Canada lay beneath the waves. The Rockies had yet to be born.

These vast interior continental seas must have been a bit peculiar as oceans go. The sea seldom reached more than 200 feet deep, and such a thin sheet of water spread over thousands of miles must have behaved quite a bit differently from what we are accustomed to seeing along our present-day coasts. Waves would have dissipated long before they reached the shores of north-central Canada. Evaporation would be high, and circulation of fresh oceanic waters with normal salinity (about 35 parts per thousand) would have a difficult time reaching the innermost regions of these epicontinental seas. At times, circulation was so impaired that huge quantities of salt precipitated directly from the seas: Michigan, always a basin, was ringed by coral reefs during the Silurian Period about 420 million years ago. The dead reefs later effectively isolated central Michigan, cutting it off from the main seaway. Evaporation greatly exceeded rainfall and what little water could seep in through the reefs; the result: thousands of feet of salts accumulated in the center of lower Michigan, to this day the prime source of table, rock and other salts in the United States.

The world was different in other ways 380 million years ago. The continents have been on the move, and North America stood astride the equator, which ran up from present-day Mexico, cutting north and east across the heart of the continent through Hudson's Bay. North America was tilted clockwise a bit back then, relative, that is, to our present orientation. And the seas that covered the land were tropical. Corals, even coral reefs, were abundant, in keeping with the modern observation that massive corals thrive only within the relatively narrow belt 30 degrees on either side of the equator. The old epicontinental waterway was warm, shallow and pretty homogeneous over a huge area.

And it teemed with life. A diverse array of shellfish, descendants of the explosion and proliferation of marine animals that had come about 200 million years earlier, covered the sea bottom, set up reef communities and exploited the available habitats. All of these invertebrates would seem at least vaguely familiar to a modern beachcomber; snails and clams, albeit of a somewhat primitive stripe, were already there, as were sponges, corals, starfish and sea urchins—all in somewhat different garb and guise than we find them in today. Instead of crabs, shrimp, crayfish and lobsters, the typical arthropods of those ancient seas were trilobites. Equipped with a pair of complex compound eyes and a set of antennae on their heads, a multisegmented, flexible body and a terminal tailpiece, trilobites scuttled around the sea bottom in profusion throughout the entire Paleozoic Era—the chunk of time that began with a rapid proliferation of all these invertebrates some 570 million years ago, and that closed with a mass extinction of perhaps as many as 90 percent of all living species at the close of the Permian Period 245 million years ago. Trilobites, brachiopods, clams, snails, corals and sponges—all these and more left their shells buried in the muds and soft calcareous oozes of the seafloor. These ancient Paleozoic seas teemed with life, and the limestones and shales that form the country rock of America's Midwest are likewise teeming with their fossils.

The Devonian Period—specifically the Middle Devonian, a span of some 13 million years that produced one of the first examples of an evolutionary pattern we interpret as "punctuated equilibria" —also saw the rise of land-living animals and plants. During the Lower Devonian, seaways were narrowly confined to just the eastern and far western margins of the continent; limy (calcareous) sediments accumulated in a long, linear trough along the Eastern Seaboard. With the advent of Middle Devonian times, all that changed—for the dramatic reason that the European continent and North America drifted together, radically altering the format of habitat and environment, the very placement of land and sea, the nature of the sediments building up in the shallow continental seas—and, as well, the very nature of life itself.

The intercontinental collision crumpled and thrust up a rim of rocky mountains from New England south down what is now the Atlantic Coast. Part of the sequence of events culminating 100 million years later that built the Appalachian mountain chain, these old Devonian mountains immediately started to erode—and as they did so, all the silt, sand and pebbles were washed down to the seas, which by then had spread over the interior of North America. As the mountains grew, great fans of sediment built up into typical deltas, and throughout the Middle Devonian a great deltaic tongue of sand and mud crept westward across New York and Pennsylvania, and over Scotland and Norway as well. As the deltas grew, they displaced marine environments, replacing them with near-shore lagoonal, riverine and lake deposits. And as the seas gave way to the rising land, so too did the brachiopods, corals and trilobites cede their space to land-dwelling counterparts: plants, insects, spiders, scorpions—and amphibians, the first of the terrestrial vertebrates.

As strange as North America back in the Devonian would seem to us now, there is a familiar look to it after all. Lands and seas have shifted around, and the precise identity of the actors on the ecological and evolutionary stage—the animals, plants, fungi and microorganisms—has changed, in some cases to a radical extent. But we

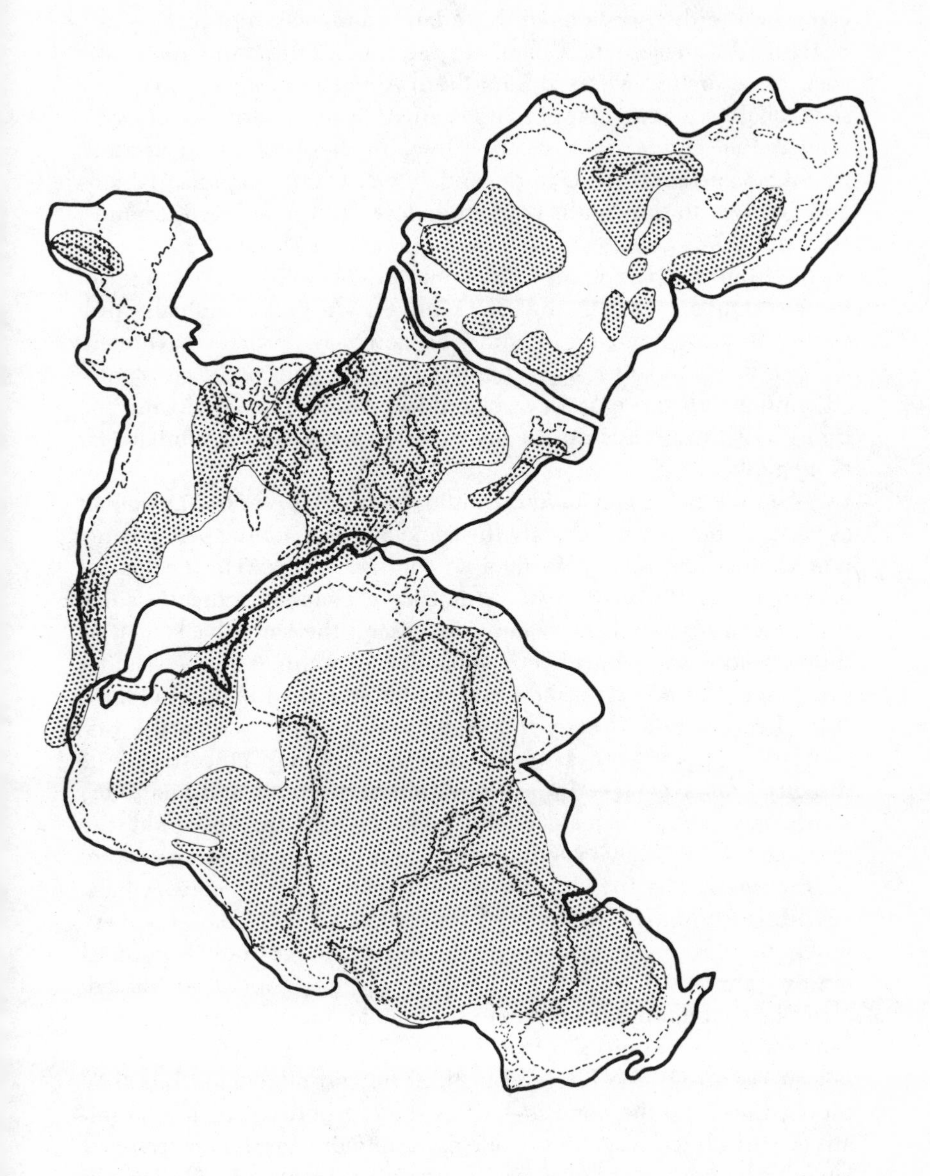

The world in Middle Devonian times. Seas are white; land is stippled.

can map the distributions of those lands and seas, and make sense of them. As we look at a map, we see a band of mountains in the east, more or less where the northern Appalachians run today. We see a shallow seaway lapping up against these eastern mountains, a seaway that extends out as far as Iowa in the west, as far north as Michigan and southern Ontario, and down as far perhaps as Alabama and Georgia to the south. Looking below the waves we find muds, silts and sands along that eastern, mountainous shore—coarse sediments from the mountains that cannot possibly reach the center of the sea. Out in what is now the Midwest we find organically produced limy muds mixed in with what few clay particles have managed to float out so far from shore. The sands and muds, constantly streaming into the eastern sea, build up thick deposits, while the limes and limy muds out in the middle of the sea accumulate far more slowly.

And we can understand the animals and plants of the Devonian as well. If they are not exactly the same as our modern species, they nonetheless are related to familiar modern forms. The ecological exigencies of life in the sea cannot have changed radically since then, and in general the clams were doing the same back then as they are now: some burrowed; some merely reclined on the seafloor (there seem to be rather more burrowers now than then). As today, some lick bacteria directly from sand grains, while others (the vast majority, perhaps then as now) strain bacteria and other nutrients directly from seawater. In fact, filtration of nutrients and finely disseminated food particles—notably bacteria—is a time-honored way that many groups have used to eat since the very beginning of complex animals. The list includes not only clams, but sponges, corals, sea lilies (crinoids—relatives of starfish still alive today, but very much more diverse and abundant in the Paleozoic), bryozoans ("moss animals") and the omnipresent brachiopods—bivalved, clamlike (but *not* mollusks) creatures that dominated most Paleozoic sea-bottom communities. Most trilobites probably roiled up the sea bottom, and they too filtered out bacteria and other food bits they mucked up from the bottom—just as some primitive crustacean relatives still do today. And, of course, there were carnivores: the first jawed fish appeared only 30 million years before the Middle Devonian began, and an assortment of squidlike relatives of the modern pearly nautilus roamed the seas and grabbed what they could. In detail, yes, both the physical habitat and the members of the biota were different from anything around today. But not so different that we cannot comprehend them using the same battery of ecological and evolutionary theory, plus observations we bring to bear from the modern world. And that is both the strategy and the ultimate value of studying things of the past.

HISTORY AND SCIENCE

Paleontology has several reputations. The rough-and-tumble field expeditions to remote and often harsh climes conjure a romantic vision. The confrontation at gunpoint between rival parties of bone hunters in the American West late in the nineteenth century fit this fantasy perfectly, and there is more than a bit of *Raiders of the Lost Ark machismo* in many a fossil collector. Yet in more pristine scientific circles, paleontology comes off more as a Victorian exercise in natural stamp collecting than as a "precise" modern science. There is a bit of truth to both imageries, of course, yet each is utterly exaggerated. It is thrilling and sometimes a bit glamorous to strike out into unfamiliar, even slightly hazardous terrain and occasionally come up with a fantastic "eureka," a beautiful specimen of some long-dead creature never before seen with human eyes. But fieldwork is, sad to say, more often dull, repetitious and unrewarding. I used to wait impatiently throughout the long winter months to get out into that ancient midwestern Devonian sea—only to yearn for my air-conditioned office back in Manhattan not long after we had hit the road. Maybe the Midwest—and my trilobites—just aren't romantic enough.

And it is also to some extent true that the elements of *doing* paleontology remain very much recognizable from Victorian times. After we have our fossilized creatures well in hand—collected in much the same ways they've always been—we still must go through the same basic steps: we must examine each specimen carefully, inside and out, to glean as much anatomical information as we can. We must decide what these fossils *are:* Are they members of a species someone else has already found and described? The literature must be searched, and we must carefully compare our newly garnered specimens with published descriptions and illustrations of species already known. We must also compare them with specimens in other collections. If an exhaustive search seems to show the newly collected material not to fit in well with any other known species, the procedure, then as now, is to make a formal description of the anatomical features (particularly the unique ones) that set the new fossils apart from all others. In so doing, a paleontologist gives the creatures a scientific name, a Latin binomial such as our own *Homo sapiens*. We, members of the species *Homo sapiens*, are in the genus *Homo* (along with the extinct species *Homo erectus* and *Homo habilis*), family Hominidae (along with species in the extinct genus *Australopithecus* and the great apes), infraorder Catarrhini (with the Old World monkeys) and the Order Primates, where all of the above, including ourselves, are united with the rest of the monkeys, plus lemurs and other "prosimians." But we are also mammals (all creatures with hair, three middle-ear bones, mammary glands), amniotes, tetrapods, vertebrates, deuterostomes, coelomates, metazoan animals. We, with plants and fungi now added, are eukaryotes: all our cells (save such specialized ones as red blood cells) have nuclei bounded by a double-walled membrane. And, with the rest of life, we are organisms. We all have RNA, and most of us have DNA (viruses do not). We, as are all species, are part of an ever-widening, increasingly all-embracing series of nested circles—and it is the job of paleontologists, and systematists who classify living creatures, to find each species and identify its position in the natural hierarchy of all living things.

Some scientists, including a goodly number of paleontologists, find this deadly-dull stuff. It is always more exciting, or at least more appealing, to be engaged on a scientific *frontier:* to be part of the latest excursion into truly new areas. I once ran across a thick tome, published in the first decade of this century, on the principles of physiology—the processes of digestion, respiration, nerve function and so forth. The author's foreword was a paean of praise to the harnessing of electricity, which made possible a magical assortment of new experimental devices that had already allowed biologists to probe far more deeply than ever before into the mysterious inner workings of organisms. No more, mused this physiologist, need we fool around with the amateurish sort of natural history that biology had followed since Darwin. Nor was he wrong to be happy about the new instrumentation which allowed him and his colleagues to forge ahead and find out still more things physiological. But if one way science advances is by the development of new techniques, new apparatus to literally let us see things finer and finer, or farther and farther away, we have merely added to our repertoire of things we think we now know and understand about the material universe. True progress in science does result in the relinquishing of older notions, and sometimes even entire avenues of inquiry, if they've actually been supplanted by a newer understanding. But some things just never go away. One of them is the natural hierarchy of life: species exist out there, more than ever in need of careful study. No manner of alternative approaches to biology have supplanted this need to understand.

This hierarchy of life—the pattern of internested resemblances interlinking all organisms—was perhaps Darwin's best argument of all that life has in fact evolved. After all (and as I have discussed in *The Monkey Business*), *if* life has evolved, we would predict that some sort of features of that primordial ancestral form would be passed along to all its descendants—all forms of life around us today. That's what we do see: RNA, and to a slightly lesser extent, DNA, are precisely that: a biochemical inheritance common to all life. Later evolutionary inventions, novelties introduced into the genealogical tree after some separation of genealogical lines had already occurred, were passed on only to descendants of one particular species. All mammal species have hair, bequeathed from the original species with hair. Hairlike structures abound in other nooks and crannies of the plant and animal worlds, but true hair is found only in mammals.

The crucial point here is that the very notion of evolution—the idea that all organisms are related, descended from a common ancestor sometime in the dim past (more than 3.5 billion years ago, the age of the oldest fossils so far found) is *testable*. The very hallmark of science, distinguishing science from other systems of thought about nature, is that whatever we choose to say about the material universe, we are constrained to put it in terms that we can evaluate simply using the evidence of our senses. We must be able to go to Mother Nature with some objective criterion to let us know if our ideas make sense.

Karl Popper, a physicist turned philosopher whose influential works began with his *The Logic of Scientific Discovery* (1934), has recently been adopted by a number of systematists, paleontologists and other evolutionary biologists. His impact has been salutary: evolutionary biology has been beset with a general propensity to take existing theory and simply use it, uncritically, to explain all manner of phenomena. All one needs is a glib, imaginative mind. Natural selection in particular has been used so carelessly that some biologists have accused their colleagues of writing "Just So" stories about how giraffes got their long necks instead of doing proper science. The charge that some aspects of evolutionary biology aren't very hard-core science comes from within as well as from without the ranks.

Popper, in a nutshell, says that a statement, to be scientific, must be "refutable." We get ideas about nature. Popper says he doesn't know how those ideas are generated—they sort of pop into your head. I think most ideas, to the extent they are not consciously or subconsciously borrowed from someone else, are subliminally suggested by patterns we sense in nature itself. But no matter, according to Popper: the crucial thing is what we do with these "conjectures." We put them in such a way that we can assess their validity when we confront nature—usually indirectly, by generating predictions about what we should find *if* the conjecture is "true." The joker in the deck is that no number of observations can absolutely prove the truth of any proposition. If we conjecture that all crows are black, and we keep looking at crows, we may indeed find only black ones in the eastern United States. But hooded crows in Europe are black-and-gray, and we have falsified the idea that "all crows are black." Popper simply tells us that we can repeatedly confirm a hypothesis by confirming all its predictions over and over again. Yet that hypothesis may someday fall to still further testing. We can speak only of "well-corroborated" hypotheses—ideas about the universe, such as the commonly accepted precept that the earth is an oblate spheroid. Or the notion that life has evolved, a testable idea whose major predictions are continually evaluated and confirmed daily by systematists, paleontologists, geneticists and those who do applied research in medicine and agriculture (see my *Monkey Business* for more on the testability of the general notion of evolution).

Some philosophers and not a few biologists have cringed at this rather astringent, acerbic, even "a-human" view of the nature of science. And it is true that science, as a human enterprise, is neither so simple nor so bloodless as my (admittedly oversimplified) rendition of Popper's thesis would have it. Popper was describing the logical structure of the way things ought to be. In the real world, in the competitive fray that is science, data forging, plagiarism and all manner of base and venal but utterly human failings make a mockery of the counterimage of detached objectivity. Such pure, dispassionate, cold logic is rare—though more common, one assumes, than the cheating of its opposite extreme. But no scientist, at least any worthy of the name, can be expected to sit back calmly and devise still more critical tests for a pet idea (though when he or she is emotionally attached to a theory it behooves our scientist to make sure that all the avenues of obvious "refutation" of the system are well understood). But scientists, as individuals, *do* argue in favor of the "truth" of this or that favored proposition—on the face of it not a very scientific mode of behavior in strict Popperian terms. Popper's critics are fond of speaking of "naive falsificationism"—boiling down simply to some biologists' stretching the criterion too far, denying that even the simplest hypotheses about evolution are "testable" in any formal sense, hence rejecting the possibility that evolution (particularly notions of *how* the evolutionary process works) can be studied at all scientifically. Such nihilism is distressing, of course; but the Popperians in biology in general are not naive, and they have done us all a service by making us sharpen up our scientific act.

Where Popper's views and the actual day-to-day workings of science coincide is in the collective effort. Science is competitive; it is, as Popper says, a collision of ideas with observations. If not all individual scientists can be paragons of disinterest in the ultimate fate of their ideas, if instead they tend to cling to favored notions sometimes in the face of rather plain evidence to the contrary (as Teggart said Darwin himself was doing), it is of no particular import. Science needs its advocates of definite points of view. It is someone else who will blow the whistle; someone who, far from committed to an idea, may just as emotionally be opposed to it—or to its proponents. It is the rivals who can be counted on to falsify an hypothesis, to claim that someone else's pet idea just doesn't square with the evidence of our senses.

BACK TO THE ROCKS

The ancient inland sea of the Middle Devonian has left a scattered record of rock outcrops, the lithified remains of all those sediments which piled up over the 8-million-year interval as the sea came and went. Stratified rocks, with layer built upon layer, older at the bottom and younger at the top, have been likened to the pages of a book: each stratum is a leaf in the book of earth history. A beguiling and not unapt analogy, perhaps; but it has led to the supposition that the earth's history—and the history of life interpreted from the fossils entombed in those strata—can be read as directly and simply as the pages of a book. We may simply climb those rocks and history will fall out before our eyes. Now, *that's* naive.

A scientist must have some idea in mind before setting out to look at nature. Even if the idea is "simply" to trace the evolutionary history of some group of organisms, such as a particular trilobite stock in this ancient Devonian sea, there must be some formal set of rules of procedure. And there should be some set of expectations, notions about what one expects to find out there. I set out to investigate the evolutionary history of the distinctive Middle Devonian trilobite *Phacops rana* for a very definite set of reasons. Most generally, it seemed obvious that the fossil record of life must have something to do with evolution—that since it presents a temporal record (however imperfect) of life's past, there must be something to be learned about evolution from a detailed study of a single species, or group of closely related species, throughout their history. Not an original thought, of course, but I was still shocked to find, once well along the way to donning the garb of a paleontologist, that my profession paid scant heed to evolutionary biology in general. Paleoecology, which uses current ecological theory to unravel ancient environments, was the most overtly biological theme in invertebrate paleontology in the 1960s. Not since the great vertebrate paleontologist George Gaylord Simpson wrote his *Tempo and Mode in Evolution* (1944), its updated version *The Major Features of Evolution* (1953) and *Principles of Animal Taxonomy* (1961) had a paleontologist truly addressed in any detailed, rigorous way the integration of evolutionary theory with patterns of evolutionary change in the fossil record. And on the whole the feeling in the profession seemed to be that Simpson had already said it all.

But some of us—and I speak here of the group of graduate students at Columbia University, mostly working under Norman Newell and Roger Batten at the American Museum of Natural History—went into paleontology because of a fascination with evolution. And we were all convinced that invertebrates—clams, snails, trilobites and other backboneless creatures which we could collect by the thousands—were at least as valuable as vertebrates for the study of evolution. Not to imply that others elsewhere weren't busily engaged in precisely the same sort of study, but I was a member of the group at the American Museum that provided the context and stimulation for the studies which produced the notion of "punctuated equilibria."

Among that core group were T. H. Waller, Harold Rollins and Stephen Jay Gould. Waller came first and was busily engaged in a detailed study of the evolutionary history of the Atlantic sea scallop, *Argopecten gibbus*, and its close collateral kin through the past 18 million years or so. Waller set the pace for the rest of us: he showed that careful selection of a group of well-preserved fossils, easily collected over a formidable chunk of time and occurring all the way from New Jersey and Maryland south through Mississippi along the Gulf Coast, could lead to a detailed evolutionary analysis. Perhaps most importantly, Waller was able to show that the very concept of species had some validity through long intervals of time. Species, he demonstrated, need not be thought of simply as mere ephemera, arguably discrete and real at any one point in time (such as in the modern world) but destined in the nature of things soon to become unrecognizably modified given the mere passage of a bit more time. Waller is now a curator in the Department of Paleontology at the Smithsonian's National Museum of Natural History.

Stephen Jay Gould, of course, codeveloped the notion of "punctuated equilibria"—and even coined the expression. That was a bit later, in 1970/71, after he had finished at Columbia and the American Museum and gone on to teach at Harvard. In the '60s he was already pursuing his multifarious interests in evolutionary biology and showing his fellow graduate students that it was altogether fitting and proper for younger members of the profession to think in theoretical terms, and to publish papers directly addressed to evolutionary theory. Gould's early research focused on what he called a "microcosm"—the half-million-year history of a land snail, *Poecilozonites*, found only on Bermuda. The fossil sand dunes and lithified soils that yield fossils of this striped, dome-shaped snail provide a detailed sample of all shapes and sizes its several species have ever manifested—and provided one of the two examples of punctuated equilibria we used in our initial paper in 1972.

H. B. (Bud) Rollins, also interested in snails, came from upstate New York, smack in the middle of the Middle Devonian. It was through Rollins that I learned how to do paleontology in the field, and in the lab as well, starting with how to find likely localities (Rollins' father-in-law, then a road commissioner in a small central New York town, knew of all the obscure little quarries opened up on farmland for crushed stone for paving roads. Informants such as he are as useful in paleontology as they are to ethnographers in more exotic climes). Through Rollins I learned such basics as knowing where you are in time when standing in a lonely cow pasture, and even how to prise specimens loose and wrap them properly. Back in the lab, cleaning fossils, removing excess gobs of rock matrix to reveal all the ins and outs of a fossil's anatomy, also requires a battery of techniques and implements, ranging from a dip in the acid bath (when impervious fossils are buried deep within soluble limestone) to meticulous scraping and picking away with a dental pick. All this before any analysis, even of the simplest sort, can be attempted.

It was Rollins who told me about *Phacops rana*. Casting around for a suitable organism for a detailed evolutionary study, I realized there were a few essentials that simply had to be met. The creatures had to be well preserved and relatively easily collected: a central tenet of prevailing evolutionary theory, a point established originally by Darwin, is that evolutionary change is based on existing patterns of variation in natural populations. It had become commonplace at least by the 1940s to recognize the importance of the study of variation, and it seemed critical that enough specimens be collected at each locality for its inherent population variation to be apparent. And, of course, as in Waller's and Gould's studies, there had to be enough *time* available for an expectation of a reasonable amount of evolutionary change to occur.

And there was yet another prerequisite: patterns of variation seen in modern organisms that seemed to point to significant evolutionary change through time more often than not were geographically based. Species may be variable in any one place, but truly significant variation often shows up only when a species is examined throughout its geographic range, where it can be found in a wider range of environmental situations. Again with a basis in Darwin, who was well aware of patterns of geographic variation and the replacement of closely related, slightly dissimilar species in disjunct geographical regions (and who used such patterns to bolster his very argument that life *has* evolved), the importance of adaptively based geographic variation had been particularly stressed by ornithologist Ernst Mayr in his *Systematics and the Origin of Species* (1942). So it seemed important at the outset that *more* than a great deal of time was necessary. I needed a subject well preserved; easily collected; with a long history, to be sure, but also spread out over a range of environments in as broad a geographic area as possible.

Enter the Middle Devonian. Remnants of that ancient seaway are preserved throughout the Appalachians and the Midwest. As we have seen, and as I soon learned, there was a decent range of habitats and ecological communities spread throughout the seaways, from the mud flats and sandy near-shore environments typical of the East to the clearer-water, limy-bottomed conditions more typical of the Midwest. And most outcrops had fossils in profusion.

Perhaps the trickiest aspect of the game of collecting over half a continent and making some sort of sense of what you find is that problem of knowing where you are in geological time whenever you stop at a stream bed, cement quarry or roadside outcrop. Paleontologists usually cannot tell time with the handy radiometrics of geochemists, who use the known rate of decay of radioactive into stable isotopes to calculate the age when a rock originally crystallized. Sedimentary rocks more than a few million years old simply do not have such isotope-bearing minerals crystallized *de novo* within them: unlike a granite or a high-grade metamorphic rock formed under tremendous heat and pressure, sedimentary rocks are simply accumulations of clay minerals, sand particles, small flakes of mica and the like—all formed earlier elsewhere and simply washed into the lake or seaway where they finally come to rest and pile up on the bottom. Some minerals, like the organically precipitated calcium carbonate that forms the basis for so much of the limestone (exploited by so many cement manufacturers throughout the Midwest), *are* formed in these sedimentary basins, but contain none of the radioactive isotopes, such as uranium and thorium, which decay so slowly to stable isotopes of lead that geochemists can measure how much of each is present and date a rock in the hundreds of millions, or even in the billions, of years.

Paleontologists cannot operate this way. There is no way simply to look at a fossil and say how old it is unless you know the age of the rocks it comes from. Sometimes igneous rocks, rocks we can date chemically, intrude into sedimentary rocks, and in such a fashion some hard-core "absolute" dates—expressed in terms of millions of years—are available for all subdivisions of geologic time. The earth is 4.55 billion years old (give or take a few million; the date comes from moon samples, meteorites and a graphic extrapolation from rocks dated directly on earth). The oldest rocks dated on earth are about 4 billion years old. We know that the Devonian Period began about 408 million years ago and ended roughly 360 million years ago. The Middle Devonian came in about 380 million years ago.

But none of that helps in a cow pasture in upstate New York. Long before radioactivity was known to physicists, paleontologists had another way to tell time. Fossils occur in the same vertical sequence throughout the geologic column. The same, or closely similar, fossils frequently occur in many far-flung localities; some are even found worldwide. This repetitive pattern of occurrence allows geologically minded paleontologists to *correlate:* rocks are mapped, and frequently certain distinctive horizons, such as volcanic ashfalls, can be traced over great distances. But rocks in isolated quarries can be matched up according to the nature of the fossils they contain. And this poses something of a problem: if we date the rocks by their fossils, how can we then turn around and talk about patterns of evolutionary change through time in the fossil record? We need an independent time frame to know that a trilobite in Ohio is roughly the same age as one in New York before we can talk about geographic variation; otherwise, their differences might as well be ascribed to the sort of process of gradual change that Darwin thought was inevitable with the simple passage of time. The distinction between the two—gradual temporal transformation of an entire species versus geographic differentiation within a species—is crucial, and indeed underlies the very notion of punctuated equilibria.

Rocks of Middle Devonian age crop out sporadically throughout the United States. Sometimes lying deep below the surface, covered with thousands of feet of younger rock and topsoil, where the Middle Devonian does reach the surface it is still usually buried beneath a carpet of forest, farmland and concrete. In the folded Appalachians, erosion has laid the rock bare to produce some natural exposures. In the Midwest, though, it is the large open-pit quarries that provide most of the random windows into the lithified remains of these ancient sediments.

In the late 1920s, G. A. Cooper wrote his doctoral dissertation at Yale, describing in far greater detail than had ever been done the layered rocks, the stratigraphy of the Middle Devonian (called the

"Hamilton Group" after the locale of Colgate University, where Cooper had done his undergraduate work). Focusing on the fossils, but also minding the precise sequence of the rocks themselves, Cooper was able to correlate the sporadic exposures across New York—and later, throughout the United States. In the region around Hamilton where the rocks are 1,500 feet thick, Cooper went up each ravine, collecting rock samples and fossils. He became so familiar with the Hamilton Group in this region that he was able to fit the far more restricted, isolated exposures elsewhere into his scheme of the general sequence of time in New York and on down the Appalachians. When, for example, a new quarry was opened up near Sylvania, Ohio, in the 1920s, the general Hamilton nature of the brachiopods, corals and trilobites was immediately obvious to all. Grace Anne Stewart described and listed many species from a distinctive alternating sequence of gray limy shales and hard limestones. The fauna of this "Silica Shale," as the sequence of rock was soon named, quickly became world-famous. The fossils are gorgeous, exquisitely preserved and extremely abundant. Today, more than 200 species have been found within the confines of this single quarry—though now the quarry is no longer worked and has become flooded.

Cooper could do more than just determine the general Hamilton age of the Silica Shale fauna—which, after all, would only be saying that the Silica Shale had formed during some interval within the entire 8-million-year history of Hamilton time. He was able to do much better than that: a number of the Silica Shale clams and brachiopods are known only from the lower portion of the Hamilton sequence of the East. And above the Silica lay a distinct unit, the Ten Mile Creek Dolomite, that bore a very distinctive coral fauna, identical to the corals of the Centerfield Formation, a widespread unit in New York that lies roughly halfway up the Hamilton sequence. Q.E.D.: the Silica Shale seemed to be equivalent in age to the "Skaneateles" part of the Hamilton of central New York—a much more precise correlation than merely calling the Silica "Hamilton-aged."

Cooper's correlation chart is a good example of a complex hypothesis. It is really a theory of time relationships among a far-flung sample of rock outcrops. Some of the particular component hypotheses of the theory have since been shown to be incorrect. Cooper didn't get it all right, and the chart is constantly being revised in some of its details as we learn more about the rocks and fossils of Hamilton age. But in general, his work has stood the test of time and subsequent investigation extremely well. And it served as the basis for my work on the trilobite *Phacops rana*.

Cooper's conclusions had special significance for me simply because he had not used the trilobite in which I was interested as a basis for his correlative scheme. Thus there was no danger of circularity: if Cooper had erred, and I subsequently compared two collections under the mistaken assumption they were equal in age, naturally the results, the evolutionary patterns I would report, would be less than fully correct. The pattern would depend very much on how correct Cooper had been about the relative ages of all these rocks. But at least there would be no problem of circularity. I wasn't going to argue that the Silica Shale was the equivalent of the Skaneateles because the trilobites are so similar (turns out they aren't—part of the clue to punctuated equilibria), and then turn around and discuss patterns of variation and evolution within these trilobites on the basis of my own time frame. No—correlations were reasonably secure, based on lots of hard work by others on different sorts of fossils. The Middle Devonian Hamilton Group and its western equivalents seemed an ideal setting for an evolutionary study of some group of invertebrate organisms.

For every vertebrate fossil, every isolated mammal tooth, as well as every *Tyrannosaurus* skull, there are thousands, probably hundreds of thousands, of brachiopod, clam and snail shells, crinoid stems, trilobite skeletons, coral calyces and all the myriad other forms of invertebrate creatures that have left some part of their hard anatomies to be trapped in the sediments and preserved in the rocks. Many of the species of the Hamilton Group, to judge from the older monographs, seemed very widespread throughout the old sea, and many of them seemed to persist pretty much through the entire 8-million-year interval of time. The problem was to pick one that seemed most promising, most likely to exhibit evolutionary change.

I was raised, professionally, to think that change *was* inevitable given an appropriately awesome span of years over which I might trace the vicissitudes of a single lineage. I bought the whole conceptual package. If there was anything I wanted to add to the picture, it was simply a comparison of patterns of geographic variation within a single species during any one brief segment of time, as compared with the long-term, protracted change I expected the entire lineage to show. I expected, in short, a slow, gradual modification of my stock—whatever it was going to be, brachiopod or clam, bryozoan or trilobite—over 8 million years, each progressive step along the way based to some extent on earlier, preexisting variation found throughout the habitats to which the groups were adapted. In other words, I expected merely to confirm a well-worn truth, a pattern everyone knew ought to be there. It seemed that the scarcity of really well-documented examples of gradual change in the paleontological literature mostly reflected a poor choice of subjects: most paleontologists interested in evolution worked on vertebrates, and the vertebrate fossil record really is, for the most part, far spottier than the invertebrate record.

There had been a flurry of work, beginning in 1899 and ending thirty years later, all centered in Darwin's homeland, that had used invertebrate fossils to show the gradual nature of the evolutionary process. A. W. Rowe collected legions of Cretaceous sea urchins from the chalk cliffs of Dover, concluding that a series of three species formed an ancestral–descendant lineage. According to Rowe, a species gradually transformed, becoming progressively better adapted to deeper burrowing behavior. In 1910, R. G. Carruthers followed suit, demonstrating what he thought was a gradual transformation within a coral lineage from the Carboniferous of Scotland. Then A. E. Trueman, in a study that has received a great deal of revision and scrutiny, wrote of the gradual increase in coiling and size of the Jurassic curved oyster *Gryphaea*. R. Brinkmann, a German working in England, ended the flurry with an analysis of evolutionary change in a Jurassic ammonite (a now-extinct relative of squid and octopi with a coiled external shell—once again proclaiming that most changes he saw occurred in the gradual, linear mode that we had all come to expect. In most such studies the basic data, the directionality of change within these various lineages of diverse ages, were correct. But their evolutionary significance was largely misinterpreted.

There have been other such studies, both earlier and later, but the time seemed ripe for some more, and Waller, Gould and I, independently, but mutually influenced, set out to provide them. There was just one other *desideratum*, one criterion for picking a suitable subject: if you believe that evolution is slow, steady and gradual, to be measured only on the gross scale of millions of years, you are well advised to look at the most complex organisms around.

In a smooth, relatively featureless shell, it will simply be harder to detect any change. Even were change inevitable, as we all took it to be, it will still be easier to see it in a creature with more parts—the additional anatomy simply providing more opportunity for nature to work its way and wreak some modification.

Trilobites fill the bill nicely. Complex, with eyes (usually) and a complicated series of ridges, furrows and surface sculpture (most of which can be interpreted satisfactorily in terms of anatomical function), and sporting an array of flexible thoracic segments and a solid tailpiece, trilobites provide a welcome wealth of anatomical detail that is simply lacking in many other invertebrate groups.

A quick search of the literature showed four fairly common trilobites in the Hamilton Group. Of these, *Phacops rana* seemed by far the most common and widespread. It even appeared to vary a bit geographically, given the mild spate of subspecific names (names alluding to geographically somewhat distinct "varieties") that had popped up in the literature. *Phacops rana* it was to be, and armed with a load of maps, chisels and hammers, and accompanied variously by wife, brother and friends, I proceeded to the next job: to sample up and down those 8 million years, and back and forth across the ancient inland Hamilton sea.

3 AT SEA IN THE AMERICAN MIDWEST

THE EUROPEAN–NORTH AMERICAN collision that began about 380 million years ago did more than change the face of the globe: it also grossly affected the face of life in North America. Many of the Hamilton species that were to dominate American life for the next 8 million years were immigrants from Europe and Africa, derived from species living in the early Middle Devonian whose fossils now come from the Rhine Valley in Germany and the desert reaches of what was, until recently, the Spanish Sahara. *Phacops africanus,* a species closely related to *P. rana,* is known from northern Africa, and it, several other trilobites and a number of brachiopods, clams and snails simply came in and took over the new habitats that were formed as the inland sea encroached over the continental interior. This "Christopher Columbus" effect was repeated several times in the history of life. The "new world" was the land of opportunity over and over again during the past 600 million years.

But I knew nothing of this. In by far the best account of *Phacops rana* that had yet appeared, J. M. Clarke (a youthful collaborator of the aging pioneering American paleontologist James Hall) remarked in 1888 that *P. rana* seemed far more like the German species *P. schlotheimi* from the Eifel district of Germany than it resembled the species from Lower and Lower Middle Devonian rocks here in North America. Conventional wisdom, of course, would see these species *(P. logani* and *P. cristata)* as the ancestors of *rana.* Clarke was right: *schlotheimi* and *rana* are virtually identical. But "identity" poses a lot of problems to a novice. Rather than thinking it interesting that two fossils so far removed in space (though plate

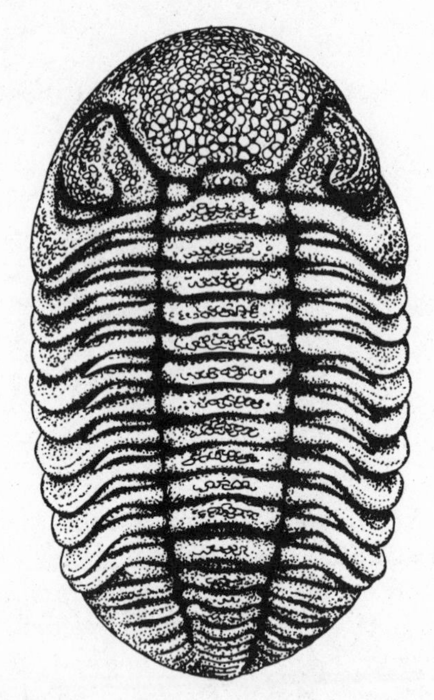

Phacops milleri crassituberculata.

tectonics *had* come in by the mid-1960s) and a bit in time *(schlotheimi* is a bit older than *rana)* were virtually identical, I chose to doubt my own ability to analyze trilobite anatomy. In the context of expected constant change, such close resemblance can be disconcerting.

And that was the problem: after I'd gotten accustomed to driving long distances, scouting out localities in pastures, stream beds, quarries and railroad cuts, some on the vaguest of directions, once my collecting settled into a routine and large numbers of specimens started turning up, my next big worry was that all the trilobites really did look the same. Some were large, some small. Some had the shell intact; others were simply the hardened mud imprints of the interior of the trilobite's exoskeleton. Some were crushed, and others had had vital parts—an eye, or the entire left side of the body—sheared off in some remote minor geologic cataclysm, or had simply eroded beyond detailed recognition through exposure to wind and water through the millennia. No matter; they all seemed more or less the same—no real source of disquiet while I was driving across New

York, perhaps, but soon to become a focus of desperation as the search broadened and the analysis deepened.

Results. Like all other humans starting out on some quest, on some project with a definite goal, scientists are determined to get results. Complicating the normal routine is the hassle of obtaining a Ph.D. A piece of doctoral research is really an apprenticeship, and the dissertation a comprehensive report that shows the candidate's ability to frame, and successfully pursue, an original piece of scientific research. Sounds reasonable, but the pressure for results, *positive* results, is enormous. If your choice is to look at evolution, and you've carefully picked out a trilobite species that meets all the criteria for a good example, and if your preliminary forays reveal a rather distressing sameness to the beasts from New York to Iowa, from the beginning of Hamilton time on up through its last gasp 8 million years later, a feeling of desperation is inevitable. For little or no change to be readily apparent over all that time and territory seemed then inconceivable—given the goals, the aspirations and, really, the basic underlying assumptions I brought to the study in the first place. Despair came full-blown late one particular afternoon in Alpena, Michigan, when, as my clothes were drying in a launderette, I took an exquisite specimen out of my pocket, pored over it with a magnifying lens and concluded it was the very same creature I had been seeing all through the Appalachians and all over the Midwest.

But launderettes, happily, are not the best of places either to study trilobite comparative anatomy or to frame profound conclusions in evolutionary theory. Part of my initial frustration *was* an inability to discern fine detail—to catch subtle differences between specimens within a single sample and, especially, to see how relatively homogeneous samples differed one from another. That Alpena specimen, it later became obvious, wasn't even a member of the *Phacops rana* lineage. It was, instead, a specimen of *P. iowensis* (which, it turns out, spent most of its time in Michigan). There was, as it later became clear, some evolutionary change in the *rana* lineage. It just wasn't very obvious, nor did it come packaged in the way I had expected. To see it I had to get everything back to the lab, cleaned up and pored over in great detail.

A CLOSER LOOK: THE EYES HAVE IT

Another recurrent fantasy of modern science is that data, huge piles of it, will somehow come to your aid and solve all your problems. Even if inconclusive, somehow a paper just looks better, feels more solid and respectable, if it is larded with vast tables of numbers and the even more unctuously soothing statistical parameters that seek to summarize the otherwise incomprehensible rawness of the observations. Literally a safety-in-numbers sort of feeling, this tendency was exacerbated when the first high-speed computers (now gone the way of the dinosaurs) became commonly available on university campuses across the nation. It seemed incredibly easy to do very slick, hypermodern science: just grab a bunch of fossils, measure the hell out of them, crank them through the IBM 7090/7094—preferably using one of the sophisticated multivariate statistical procedures—and *voilà!* instant answers, instant results, instant success.

Not to unduly disparage computers or statistical analysis in paleontology or any other discipline. Whatever drawbacks they pose lie merely in the snare-and-delusion realm: there has been (and still is) a tendency (ineluctable in some quarters) to let someone else's algorithm (numerical procedure) massage your data (more often than not these days collected by a technician) as a substitute for careful thinking about either the data themselves or even the assumptions and apparent results of the computer analysis. When used instead to help test a hypothesis, and used with due circumspection, computers of course are an enormous boon in the endless task of simplifying our observations of nature to the point where we can hope to grasp something of its complexities.

A true reflection of my cultural milieu, I turned automatically to the computer with the explicit feeling that surely *here* was a way to detect some evolutionary change—change mere eyeballing of assorted specimens to date had failed to deliver. Wrong again: I measured some 50 different lengths and widths—length of the head, distance between the eyes, height of the eyes, length of the tail and so on—on hundreds of specimens. This was tedious. Each specimen had to be cleaned at least well enough to make all the anatomical landmarks visible. Each had to be mounted on a block of wood, stuck to a blob of plastilene, with the tops of the eyes (a flat surface) in a horizontal plane, perpendicular to my line of sight down the barrel of the microscope. There was a little scale inside the right eyepiece reticle from which to read off the various measurements. After a day of measuring, and until I got used to the microscope, I would see double from the bus window on my way home.

And I used a battery of univariate, bivariate and multivariate statistical techniques to show me how these trilobites had evolved. The multivariate techniques (such as "factor analysis") perform the illusion of taking *all* measurements simultaneously, effectively letting you know how close specimen A is to specimen B, or Group A to Group B, "all things taken together." And there actually were some results. I found that the eyes of these trilobites grew a bit faster than the rest of the head early on in life, only to slow down later in life. And the eyes of one of the species of the *rana* group (there turned out to have been three main species hidden in "*Phacops rana*") got relatively bigger (relative to the rest of the head, that is) in a seemingly gradual way through a roughly 5-million-year period. But the trend was too subtle ever to see, even in hindsight, with the human eye. Otherwise, growth (from baby to adult) of these trilobites was as simple and linear as that of a clam, maybe simpler, and the computer showed nothing much whatsoever going on through time. Hence my bias: computers can be a snare and delusion to anyone banking the success of his doctoral research on the magic elixir of modern technology.

But what all that staring through a microscope, measuring all those beasts, did accomplish was get me to look at those trilobites really hard. After a while, each specimen becomes unique: its preservation, or slight distortion, or the rock matrix or other fossils with it first makes each one distinct. But then truly idiosyncratic features of each one start to pop out—especially the anomalous ones, those bitten and healed, or diseased, or the victim of some developmental anomaly. I still think the tiny little one with the growth in its left axial furrow had what must be among the earliest known tumors, at least in arthropods, if not for all organisms. (I never published that, though I was tempted at the time to fire off a note to the prestigious *Science* magazine, which, then as now, seems to dote on such stuff).

But the real key lay in the work, done in the early 1960s but not published until 1965, of a Scottish paleontologist, Euan N. K. Clarkson. Clarkson (who has been teaching since the '60s in Edinburgh) worked out the details of the visual anatomy of several species of British Silurian trilobites, all closely enough related to *Phacops* to have the same basic sort of peculiar eye. Even the computer showed a partiality to eye measurements. It was through an apish copying of some of Clarkson's data-gathering on eye anatomy that evolutionary patterns—patterns of change, and patterns of utter stability—finally became clear in my own specimens.

Most trilobite eyes look a lot like the close-up scanning-electron-microscope photographs of insect eyes that have appeared all over the place during the last ten years: a bulging surface consisting of anywhere from tens to hundreds of lenses, arrayed like the cells in a honeycomb, with adjacent lenses offset to maximize packing, and covered over with a single thin translucent corneal membrane. But the eyes of *Phacops* and its closer relatives are something a bit different from the usual arthropod affair. The lenses are extraordinarily large, ellipsoidal balls of the clear mineral calcite (calcium carbonate) and *each* lens has its own separate cornea. It is more an aggregation of a number of separate eyes than a single compound eye typical of most insects, crustaceans—and most other trilobites.

What Clarkson, the physicist/trilobite *aficionado* Riccardo Levi-Setti and some other paleontologists have managed to discover about the physiology of vision of these creatures is astounding. These visual systems, with seemingly no real close analogues among living arthropods, once again confirm the ages-old suspicion that nature truly is grand in her inventiveness—as are clever humans who can unravel such anatomical mysteries from 400-million-year-old fossils. Clarkson did a lot of careful sectioning of trilobite eyes in the '60s, making thin slices through them and studying them under the microscope. The scanning electron microscope, which provides high-resolution, high-powered magnification of three dimensional objects, was just coming into general use in the late 1960s, and Clarkson soon turned to this new tool to augment his anatomical probing. He was ultimately able to determine the true nature and shape of the lens structure of phacopid trilobites—including a reconstruction of the soft tissue around and below the hard mineralized region of the lens.

Then came Levi-Setti, who made an inspired mental link between Clarkson's photos and drawings and some illustrations he had come across long before in his work in optical physics. It took some digging, but he came up with them: in the seventeenth century, independently, Christian Huygens and René Descartes (the same

Descartes who gave us the idea that the universe is in constant motion) mathematically solved the problem of how to construct a lens that would allow light rays to pass through without producing the distortion known as "spherical aberration." The two men produced slightly different solutions to the problem, though each was an "aspherical, aplanatic" lens—that is to say, an ellipsoid. Clarkson had found two variant versions of phacopid lenses in his survey of eye anatomy. Levi-Setti simply showed that some phacopids utilized Huygens's design, while others followed Descartes's. There's at least one moral in there somewhere.

Paleontologist K. Towe, meanwhile, knew that calcite, in its clear form ("Iceland spar"), could transmit light along its three mutually perpendicular crystallographic axes. But looking through either of its first two directions produced "double refraction"—the image was split into two overlapping pictures. Hardly the stuff of clear vision. So it is not surprising that Towe found nature once again clever enough to handle the problem: each lens of a phacopid eye has the third crystallographic axis, the only one that does *not* split the image, pointed resolutely straight out. None of his images were doubly refracted—they were just inverted, as in a camera, and there are simple neurological corrections for *that*.

Why could these bottom-dwelling trilobites so long ago see so well? Why did they *need* to see so well? Clarkson was able to show by a series of meticulous measurements that not only was the clarity of their vision surprisingly good, but many of them could see a full 360 degrees around them—the visual fields of the two eyes overlapped front and back. Other phacopids had more modest ranges; but all seemed able to take a lot in, and to see it rather well. Predator avoidance (rather than capture—phacopids were not active carnivores) seems the simplest guess, but remains just that, a guess. The problem is of more than idle interest, though, as much of the change phacopids underwent in their 130-million-year sojourn between the Lower Ordovician and Upper Devonian involved modification of the eyes: their position on the head, their size, their number of lenses and so forth. And within my *Phacops rana* lineage it turns out that changes in the arrangement of the lenses in the eye was one of the key elements of their 8-million-year evolutionary history. It would be nice to know just what their vision meant to them.

Paleontologists at the turn of the century had already devoted a good deal of attention to these prodigious eyes, and the same J. M.

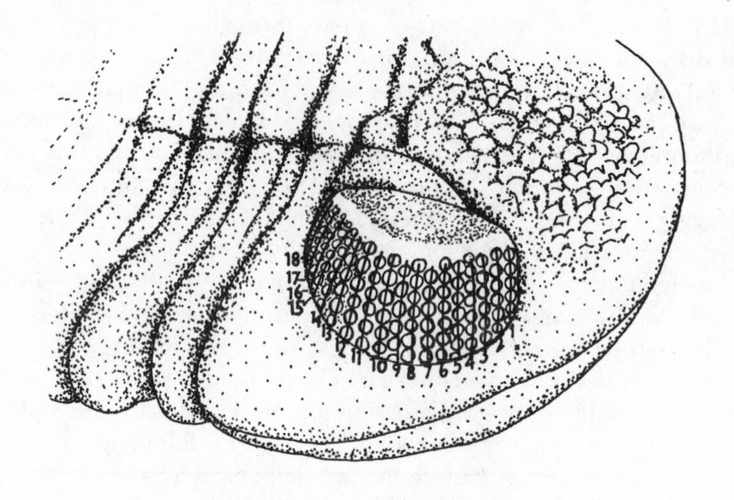

Counting lenses in *Phacops.*

Clarke had devised a system for counting the lenses of the eyes—a requisite step if any two eyes are to be compared. And here is a fateful choice: looking squarely at the curved surface of a phacopid eye, one can see those honeycomb lensar arrays in either of two ways. One is to perceive them as two intersecting sets of curvilinear strings of lenses—much like the two opposing swirls of seeds in a mature sunflower, or the intersecting arrays of an actual honeycomb. The mathematics of these lines follows the Fibonacci series, in which each number is the sum of the two preceding it: 1,1,2,3,5,8,13,21. . . . Clarke looked at the eyes and simply saw them this way. He reported that some trilobites would have 8 and some as many as 11 slanted, curved rows of lenses. There seemed no definite pattern of variation in the distribution of such curved rows either within populations or through time within a single stock—and Clarke was working mostly with *Phacops rana.*

Ignorant of the (rather obscure) paper Clarke had written on ocular anatomy in 1889, I simply adopted Clarkson's alternative way of seeing those lenses. For just as obvious as the two intersecting arrays of curved lines of lenses is the simple series of vertical columns of lenses, arrayed like the threads of a simply woven fabric. Clarkson had hit upon—and decided to retain when he published his results—a clear and simple convention of notation for his lenses. Counting from the front of an eye, and writing the lens counts for each column in clusters of three for legibility, he might find 345 656 767 655 543 21 (17, 80). This array (not atypical for *Phacops rana*) is the front-to-back count in one eye of all the lenses in each vertical column. The numbers in parentheses following the tabulated count are the number of columns counted, followed by the total number of lenses in the eye. I readily admit that I counted lenses only on the chance that something interesting would crop up—it somehow seemed the thing to do following Clarkson's work. And adopting *his* way—Clarkson's, not Clarke's—of seeing those eyes turned out to be the key to the whole analysis.

Whenever we went to a quarry or roadside outcrop, we would try to collect specimens from as thin a layer as possible—ideally 30 or more trilobites (or at least 25 to 30 heads, complete specimens not being too overly common). Optimally, all would come from a single bedding plane, an instant in the geological past. More commonly we would scour a 10-foot-thick chunk of rock, sometimes even more, to accumulate a sample. The distinction is critical: the thinner the interval, the more likely one is sampling something approaching an authentic biological population. And some of the better samples—collections made along the shore of Lake Erie, or mined from one level in the Silica Shale—quickly revealed some solid, invariant facts about growth and development of these particular trilobite eyes. Trilobites, like crabs and shrimp, shed their outer skeletons to grow. Just like a Maryland blue crab, trilobites periodically molted, quickly swelled up with a big intake of water and then quickly deposited limy salts once again to harden their outer covering. This mode of growth is a boon to a paleontologist, as it means that a single animal may leave fossilized remains as many as twenty different times during its life. Even if we cannot trace the growth of a single organism (there being no way to match up earlier and later molt stages of the same creature), we can nonetheless see statistically what the population norms of individual growth were really like.

Baby *Phacops rana* usually have a formula such as 122 323 222 211 1 (13, 24). Both summary numbers quickly leap forward through immediately successive molts. But then they begin to level off. As soon as 17 columns of lenses are reached in most samples of *P. rana* from New York State, for example, that's it—the trilobites continue growing, doubling in size several times, in fact, but the number of columns of lenses remains generally constant, left and right eye, and from specimen to specimen, stable at 17. The number of lenses keeps climbing awhile longer, but then it too levels off—and in some of the very largest specimens the total number of lenses even appears to decline, just as very old, very large female horseshoe crabs begin to resorb some of their lenses between some of their final molt stages.

One day I simply made a summary list of these lensar statistics for all the localities for which I had good samples already measured and counted. I did it on a whim, more to relieve the boredom of reaching for still another sample to clean, measure and count than out of any conviction that I was on the brink of anything exciting.

But there *was* a pattern there, one I hadn't noticed before. An innocuous list of, for example, Windom Shale, Jaycox Run, New York: 17; Menteth limestone, Jaycox Run, N.Y.: 17; Moscow Formation, Earlville State Lands, N.Y.: 17; Silica Shale, Ohio: 18; Four Mile Dam limestone, Michigan: 17. And so forth. I saw the numbers, but I could see them mapped three-dimensionally, over the inland sea and placed in appropriate positions in time. I was seeing variation between samples, and for the first time there seemed to be some simple sense to it. There is always a thrill of excitement that surges as months of frustration are resolved when the answer finally jumps out—or in this case, just some positive evidence that any evolutionary change at all had occurred in these trilobites. Salvation! There would be something to write about after all.

ONE PATTERN, TWO THEORIES

Once I had identified, finally, where at least some change had happened in my *Phacops* lineage, it was simple to survey the rest of my samples, from the oldest to the youngest, from New York out to Iowa, to see if the story held up, and to fill in the gaps in the data as they had so far accumulated. Within a few days the overall picture had jelled—and it has not changed as I and others have routinely checked all new samples that have come to light over the intervening years. And the picture that emerged is utterly typical of the way evolutionary change leaves its mark in the fossil record.

Middle Devonian rocks crop out rather more sporadically in the eastern Midwest than in New York and the Appalachian states to the south. Each exposure in Ohio, Michigan, Illinois, Iowa, Indiana and southern Ontario reveals only a fraction of total Hamilton time—but the quality of the fossils and their abundance help make up for the frustrations of spotty timekeeping in the rocks. At several sites—notably in northwestern Ohio, along the Ausable River in the Arkona region of Ontario and along the shores of Lake Huron on southern Michigan's northeastern coast—exquisite fossils, including trilobites, abound.

The Silica Shale in Ohio and the Arkona Shale in Ontario, in particular, have much to tell us about evolution. Each is a sequence of limy muds now hardened into shales, and interbedded with pure, dense limestones—quarried in Ohio for cement manufacture. Both the Silica and Arkona formations harbor two variant versions of *Phacops*, traditionally called *P. rana milleri* and *P.r. crassituberculata*. The two forms differ mostly in the number of lenses in the eye: *milleri* has so many lenses (typically over 125 in mature specimens) crowded onto its visual surface that the entire eye looks swollen. The variant *crassituberculata* has a more modest complement of lenses (typically about 85–90 in the adult), and the usual hexagonal skeletal rim is easily seen separating each lens from its neighbors. Otherwise the two hardly differ. They have almost never been found together (I know of a single instance)—nicely eliminating the possibility they are males and females. *Milleri* occurs in the shaly layers, while *crassituberculata* is invariably in the harder limestones. The early developmental stages of the two appear to be identical, and perhaps the two forms are mere developmental variants, the *milleri*-type eye having developed in the muddier substrate and the *crassituberculata* having formed in clearer waters over more purely calcareous seafloors. For reasons that will soon be more clear, the best hypothesis is to see the two as distinct subspecies of the single species *Phacops milleri*. Their cumbersome names then become *Phacops milleri milleri* and *Phacops milleri crassituberculata*—though both very much remain members of the *Phacops rana* evolutionary lineage. In any case, the fluctuation in the average number of lenses in the eyes of these rather stunning trilobites was not the focus of evolutionary change: both *milleri* and *crassituberculata* have 18 columns of lenses in their eyes.

Collecting in the north quarry of the Medusa Portland Cement Company near Sylvania, Ohio, or along the banks of the Ausable River near Arkona, Ontario, while rich and rewarding far beyond many a paleontological experience, is nonetheless far from perfect. The most fossiliferous of rocks have their barren zones, and their little pockets loaded with fossils. (When the Medusa company closed its north quarry, it reopened the old south quarry just across the road. Collectors were chagrined that the very same Silica Shale that had provided such treasures for 50 years across the street held far fewer species, in much-reduced numbers. Once again we learn that "the same rock"—sediments accumulated nearby at the same time, but reflecting slightly different environmental conditions—will harbor a vastly different fossil record.) Even the best of spots produces heads, tails, and the occasional complete trilobite only sporadically. Collecting upward through tens of feet of interbedded limestones and shales, measuring where you are from the quarry floor and identifying the different beds (referring to whatever detailed studies of the rocks may already have been published), you record the location of each fossil; but there is no millimeter-by-millimeter sampling. Serendipity in discovery and the vagaries of preservation and exposure ensure a far spottier sampling up and down the rocks.

And spotty sampling up the rock column—where tens of feet may go by without yielding so much as a single head—ensures a spotty sampling of *time*. Even a relatively thick and densely fossiliferous unit like the Silica Shale directly records only some portion of the time interval that elapsed between its start, as the first particles began to accumulate, and its end, when the uppermost layers were formed. We might judge, as nearly as possible, that it took a million years, say, from beginning to end to form the layers we today call the Silica Formation, but we have no idea, when climbing over the famous Unit 9 (a shale unit, the ninth distinctive layer from the bottom of the entire formation, and one that has yielded huge numbers of *Phacops milleri*), *when*, in that million years, Unit 9 began accumulating on that ancient seafloor.

So we are randomly sampling sporadic bits of time—and evolution—as we collect up and down these quarry walls and bluffs along riverbanks. Seemingly less than ideal, it is the best we can do—and we must wonder what we might hope to learn in such circumstances. It turns out that this time and specimen gappiness of the record really has been the basis for rationalizing away what the fossil record seems so clearly to set forth—just as Teggart was saying in 1925.

This is what evolutionists ever since Darwin have been saying about what we might expect to find *if the fossil record were complete.* The words happen to be Ernst Mayr's, but similar thoughts have been penned by many a paleontologist and evolutionary biologist:

> Hitherto we have spoken only of the delimitation of contemporary (synchronic) species. The delimitation of species which do not belong to the same level (allochronic species) is difficult. In fact, it would be completely impossible if the fossil record were complete. The species of each period are the descendants of the species of the previous period and the ancestors of those of the next period. The change is slight and gradual and should, at least theoretically, not permit the delimitation of definite species. In practice, the fossil record is fragmentary, and the gaps in our knowledge make convenient gaps between the "species" [Mayr, 1942, p. 153].

A. J. Cain, in his useful and lively book on species *Animal Species and Their Evolution,* expresses much the same sentiment: Thank goodness the fossil record is *not* complete! This is rather an odd state of affairs—expressing gratitude that our data, our information about the material world, are not better than they are. It speaks volumes about what these biologists expected to see if the fossil record were a more complete and faithful archive of time and evolutionary events. What they expected, what they predicted, was exactly what Darwin himself was looking for: the virtually inevitable drift, the gradual transformation, of a species as you collect it bit by bit up through the layers of time. And that was, of course, what I was looking for too.

But that's not what's there in the Silica Formation, the Arkona Formation or any other rock unit in the Midwest. Under the Silica Shale there is the Dundee limestone—with the *crassituberculata* form with its 18 columns of lenses. All through the Silica, all through the Arkona, whenever a specimen comes to hand, it has those 18 columns of lenses—no more, no fewer, be it a *milleri* or a *crassitu-*

berculata, and provided it isn't a tiny infant. The record is spotty, yes, but roughly the lower half of Hamilton time recorded in the Midwest yields only members of the *rana* lineage with 18 columns of lenses.

The Silica is overlain by a thin unit called the Ten Mile Creek Dolomite, while the Arkona is succeeded by the Hungry Hollow Formation, some 3 feet of shale followed by 3 feet of dense limestone. Both these formations contain the distinctive coral fauna of the Centerfield Formation of New York—that marker horizon which appears about midway through the Hamilton Group. Suddenly, all of the *Phacops* from these latter units have only 17 columns of lenses.

Now, in the entire 8 million years of Hamilton time, the greatest (though not the sole) amount of modification wrought by evolution in the *Phacops rana* stock was the net reduction from 18 to 15 columns of lenses. Hardly prodigious, this degree of anatomical retooling falls well within the normal bounds of "microevolution"—loosely speaking, the kind and degree of relatively minor change that marks the difference between closely related species, and the sort of change that can be seen in rudimentary form within a single variable species. The internal, within-species variation is then supposed to supply the raw stuff for the differences we see between species—and ultimately on up through genera, families and the really larger groups of organisms.

But, at least in the Midwest where parts of the evolutionary story of the lenses first began to come clear, we see something out of whack with prevailing expectations—two things, really. We have, it is true, a good but far-from-perfect record, and a less-than-perfect sampling of what really is there. But as we climb up those rocks and check those samples, over what must be, in sum total, a 3-or-4-million-year period, we see some oscillation, some variation, back and forth (the two subspecies coming and going with shifting substrate)—but no real net change at all, *and no change especially in the anatomical feature, those columns of lenses in the eyes, which end up showing the greatest amount of change within the entire lineage.* This is the first element: simple lack of change. Stability, or *stasis*, as Gould and I began to call it.

And the second element in this midwestern pattern is the apparent suddenness of the change: when it does come, evolutionary modification seems to be abrupt, an all-or-nothing sort of affair. This sort of jump, or apparent gap between ancestor and descendant, depending upon how you look at it, was itself nothing new. Even the historian Teggart realized that the fossil record was full of examples of episodic, fits-and-starts change. And there were, in the 1960s, two well-entrenched, opposed ways of making sense of such data, two ways of looking at the evolutionary process which yielded utterly different conclusions over the significance of such jumpy data in the fossil record. Each viewpoint has its merits, and each its faults as a satisfactory explanation for our midwestern pattern. And both, at least in this instance, turn out at bottom to be fundamentally wrong.

The competing hypotheses diverge at the outset over the simple question of attitude: how literally do we take the fossil record? The conventional view, the one that Darwin forged and we have all inherited, sees the jumpy pattern of change as an artifact. What is there is a pseudoevent. What is really there is a gap, like a blank section on a tape. The phenomenon of sudden evolutionary change is *not* real. We have two kinds of trilobites, perhaps an ancestral species and its descendant. They lie one above the other in immediately adjacent rocks. But, as Darwinian theory would predict, there *must* be time missing between the two rock units housing the ancestor and the descendant. Remember that this pattern is general: the jump between 18 and 17 columns of lenses may seem small enough, sufficiently trivial *not* to require, in theory, an inordinate amount of time. But, as a rule, most differences between apparent ancestors and descendants come this way, in what seem like sudden changes in state, from an ancestral right into the descendant form. As if it happened overnight. Conventional neo-Darwinian theory would naturally predict a temporal gap in the record—even if it were not obvious at first glance at the rocks themselves.

And this prediction of a temporal gap is nicely fulfilled in the history of these midwestern trilobites. In G. A. Cooper's reckoning of how the rocks of the Midwest fit into the Hamilton time scheme of things, both the Silica and Arkona formations seemed to have finished accumulating long before the Centerfield Sea spread over the continental interior. In fact, to Cooper and many other geologists puzzling out the geometry of rock and time in the Devonian, it seemed as if the sea had disappeared entirely in the Midwest for a good long period of time—perhaps even a million years—before it swept in once again over the continent, and the Centerfield limestone and all its equivalents to the west started to form. The Darwinians are simply right about the time gap.

The convenient thing about gaps in the record is that we need not invoke truly instantaneous, overnight evolutionary leaps to explain the transitions we seem to see in our fossils. With perhaps as much as a million years missing, and certainly for such modest change as the column counts in these trilobite eyes, we can easily maintain that evolution is, after all, a gradual, intergradational affair. We might regret not being able to "see" that transition—the gap in preservation unfortunately seeming to occur when the evolution did. But then there is that saving grace: the rather horrific thought that were there no such gaps, we simply would not be able to classify all these fossils, placing them into a neatly ordered set of internested boxes the way we sort out all living creatures.

Not so, according to a minority, but seemingly omnipresent, group of paleontologists who, again ever since Darwin, would like the fossil record, with all its different patterns of change, to be read a bit more literally—and to be taken, perhaps, a bit more seriously. There have always been paleontologists, even before Darwin, who recognized the common pattern of episodic change the fossil record so loudly and consistently proclaims. The old catastrophists, for example, with Baron Georges Cuvier (1769–1832) in the lead, were convinced that new groups were created to replace their forerunners, creatures that had lived for a spell until they were wiped out in some ancient cataclysmic extinction event. This wasn't a bad description of the way the record really looks, and after Darwin established evolution as both the genealogical interconnector and the force of change, paleontologists still attracted to the literal patterns of change they saw in the rocks became, collectively, "saltationists." "You have loaded yourself with an unnecessary difficulty in adopting 'Natura non facit saltum' so unreservedly" were Huxley's words to Darwin. Saltationists think evolution really does proceed by leaps, sudden jumps from one condition to another.

Perhaps the best-known saltationist of recent times was the geneticist Richard Goldschmidt, who published his *The Material Basis of Evolution* in 1940—just as the primary documents that served to found the Modern Synthesis were in the midst of production. Theodosius Dobzhansky had published his first edition of *Genetics and the Origin of Species* in 1937, and his second edition (with many of his additions coming as rather strong attacks on Goldschmidt) was to be published in 1941. And Mayr's *Systematics and the Origin of Species* (1942) and Simpson's *Tempo and Mode in Evolution* were likewise in the midst of production.* All three authors vigorously opposed Goldschmidt's view that the stuff of microevolution—natural selection working on a field of genetic variation, whose ultimate source is mutation—could produce the sorts of differences we see *between* species. Goldschmidt spent many years working on the genetics of the gypsy moth *Lymantria dispar*. He thought his evidence showed that patterns of genetic variation within species involved characteristics different from those which form the differences *between* species—a thesis hotly denied by Dobzhansky and Mayr. Goldschmidt imagined mutations of large effect—his "macromutations"—to be the spontaneous force behind the appearance of entirely novel anatomical packages, new sorts of organisms. His "hopeful monsters"—rare mutant forms spontaneously produced from a "normal" parent—were to him the more likely candidates for the production of the truly new forms in evolution. Most mutations, especially the sorts with large effects on an embryo, disrupt the normal course of development of the fertilized egg—usually with lethal, or at least what geneticists call "deleterious," effects. Goldschmidt knew all that, but speculated that sometimes some of these large-scale mutations may *not* be lethal, and may produce a fairly healthy if otherwise abnormal (*vis-à-vis* the population norm, that is) organism, and that that organism may just hang on, survive and reproduce. Rare, fortuitous—but, when it happens, instantaneous evolutionary change was Goldschmidt's vision of macroevolution.

The antithesis of Darwinian gradualism, Goldschmidt's notion fared poorly, posed as it was amidst the emerging orthodoxy of the modern synthesis. And the neo-Darwinists had some good points: there seemed to be no real experimental, laboratory evidence that mutations of the sort Goldschmidt fancied ever occurred. And even were such a monster miraculously to be born and survive to come into reproductive maturity, with whom would it mate? Theoretical genetics and real-world experience combined to load the argument rather strongly against Goldschmidt's version of the evolutionary

* See my *Unfinished Synthesis* for a detailed technical analysis of the contents of these major works of the evolutionary synthesis.

process—so much so that his saltationist leanings were still very much the object of scorn in the 1960s.

In the rather more uncertain world in which evolutionary biology finds itself nowadays, Goldschmidt has become a bit more respectable. His large-scale "macromutations," for one thing, no longer seem utterly groundless fantasies. Molecular biologists now realize that the genes which code for proteins are only a portion of the genetic apparatus. These "structural" genes are themselves regulated—switched on and off—by other genes which themselves do not code for such easily detected substances as enzymes. And a relatively slight alteration of such a "regulatory" gene may well have cascading effects: if a mutation in a regulatory gene causes it to switch on some structural genes relatively early in the course of the development of an embryo, for example, by the time the organism is fully developed some truly large-scale changes are bound to have accumulated. It is not the mutation *per se* that is "large"—it is the final outcome in the fully developed organism, the result of a timing change that starts the developmental ball rolling earlier in ontogeny, producing large-scale effects in the adult organism.

But Goldschmidt was a voice in the wilderness among geneticists in the 1940s, and his ideas even now lack the rich empirical support we would all like before thoroughly embracing a theory. But in paleontology, similar notions—all centering on the abrupt appearance of new structures—have always enjoyed fuller, if not majority, acceptance. And this too is a matter of empirics: the fossil record is full of examples very much like our midwestern trilobite case. Otto Schindewolf, a well-known and influential German paleontologist, recognized these patterns and insisted they be taken literally for what they seem to be: relatively abrupt evolutionary events. Schindewolf was a paleontological saltationist, and saw no reason why the fossil record should not be read literally. His attitude, at the very least, is positive: The fossil record has something to tell us directly about the nature of the evolutionary process. It is not merely an embarrassment, a document so faulty we must turn from it, explaining its pattern away as a mere artifact of gaps.

Schindewolf was a sophisticated geologist and knew that the fossil record is riddled with temporal gaps. But the cases were so many that he stuck to his guns and insisted that evolution is not the regular, gradual affair the majority of his colleagues, especially in the United States, were prone to accept. But Schindewolf, too, was struck with the stability that seems to imbue most fossil species. He spoke of "typostrophism," the sudden change from one "type" to another, as if each "type" itself were invariant, or at least extremely stable. When things *do* change, Schindewolf was saying, it is as if they were bursting out of one straitjacket and jumping right into another. *How* such change was to occur, though, was not Schindewolf's strong suit, and the majority of opinion was as much (though perhaps not quite so sharply) against him as it was opposed to the hopeful monsterdom of Goldschmidt.

Taking stock, and thinking about the pattern of change in the midwestern phacopid trilobites, there is something to be said for both the saltational and gradualistic explanations for the change. Neo-Darwinian gradualists would predict a time gap between ancestor and descendant, and, of course, they'd be right. Score one for convention. Saltationists, however, justly point to the prodigious stability they see in the ancestors, and ask an interesting question: if gradualism is the rule, why don't we see any hint of change? We might not be faced with a perfect record, but *if* gradualism is the rule, our sporadic sampling up and down cliff faces should give us some hints, some directional drifting, from the primitive state of the ancestor on over toward the condition we eventually find in the descendant. Why is *all* the gradual change going on in those very gaps?

So, here's a bit of a dilemma. When we finally find some evolutionary change, however slight it may seem, the "typostrophic" sort of affair the *Phacops rana* lineage seems to show in the Midwest poses a choice between two unappetizing alternatives: either you stick to conventional theory despite the rather poor fit of the fossils, or you focus on the empirics and say that saltation looks like a reasonable model of the evolutionary process—in which case you must embrace a set of rather dubious biological propositions. Paleontologists are rather well known for taking that latter course—adopting *ad hoc,* outmoded and sometimes downright mystical ideas about biological processes just because they fancy that these ideas fit what they think they see in the fossil record. I had every desire to avoid that well-trodden path. Besides, I was (and remain) too much of a conventional neo-Darwinian ever to subscribe to the saltationist heresy.

But, on the other hand, I was not prepared for the inertial stolidity of my fossils. They didn't seem to *want* to change. Now that

An evolutionary enigma in the Midwest.

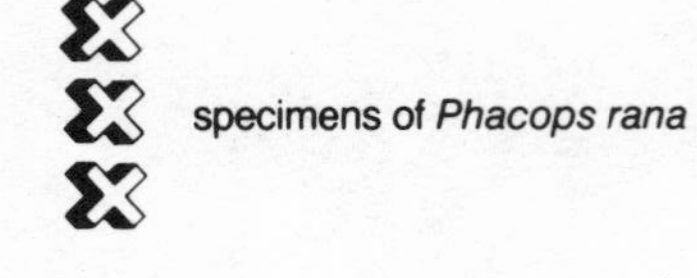

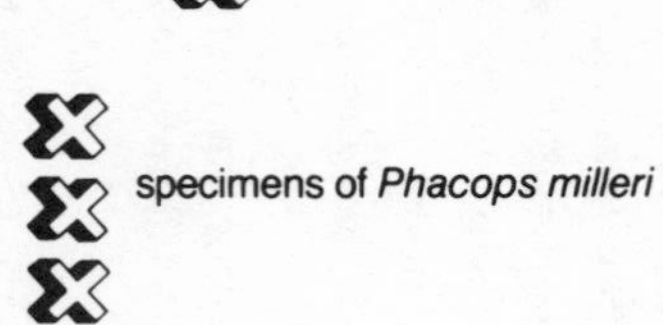

X marks the spot: specimens of *P. rana* and *P. milleri* are collected at odd intervals through local rock outcrops in Ohio, Michigan and Ontario

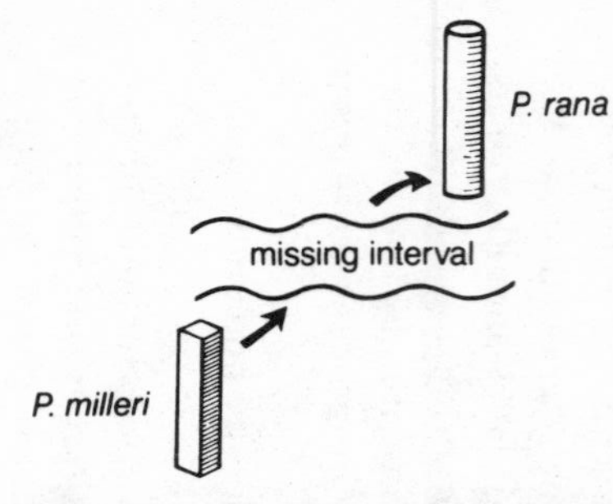

Interpretation No. 1. Standard evolutionary theory sees the evolutionary change as gradual and unrecorded in the rocks—correctly leading to the prediction that there is a time gap between the ancestral and descendant samples. The time gap *must* be there if evolution is gradual.

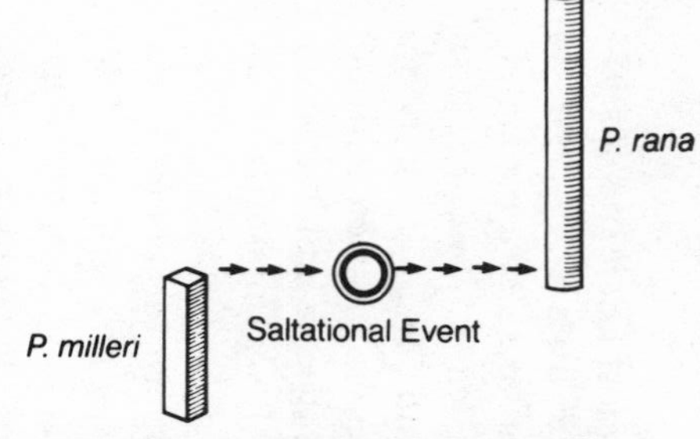

Interpretation No. 2. Saltationists would insist that the rock record be read literally. Thus evolution from *P. milleri* to *P. rana* was *not* slow and gradual, but relatively abrupt. Though it ignores the very real time gap, the virtue of this interpretation is its simple recognition of the conservative stability in eye anatomy of both parent and daughter species.

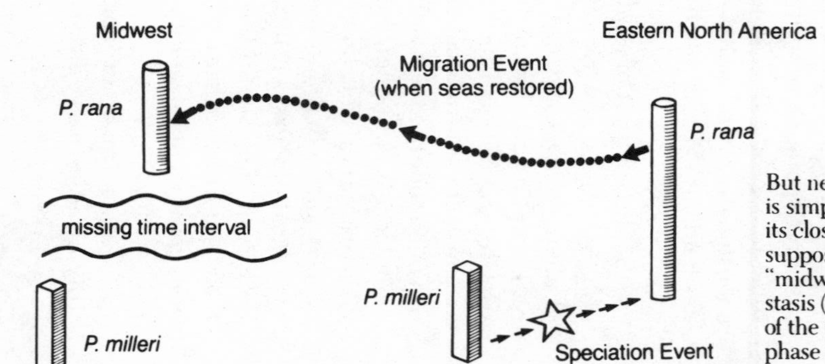

But neither interpretation is correct; what happened in the midwest is simply a biogeographic event, where one trilobite species replaces its close relative after an interval when the seas had dried up. Many supposed cases of evolution in the fossil record are events of this "midwestern" type. Thus speciation coupled with a recognition of stasis (= "punctuated equilibria") seems a better picture than either of the two older evolutionary models as a basis for understanding this phase of the evolution of *Phacops*.

some, albeit rather trivial-seeming, change had shown up (and I could relax—*something* had happened), the stability no longer seemed such a personal threat. And the stability, the lack of change, began to seem awesome. I realized it wasn't just *Phacops rana*, either. A snail, one *Bembexia sulcomarginata*, also shows up in the earliest Hamilton beds in central New York. Unlike most of the other 200-odd Hamilton species, this snail lasted only for the first 3 or 4 million years of Hamilton time. When we find it in fine-grained, near-shore muddy bottom deposits, the shell is delicately ornamented; in coarser-grained sandy bottoms it tends to be correspondingly more coarsely ornamented—more variation, more than likely reflecting adaptation to local environments. At any one time the species is variable, its variation in concert with variation in local bottom conditions. Through time, over those 3 or 4 million years, environments oscillated between sandy and muddy conditions as you climb the ravines in central New York. At the end, when last we find the snail, it once again shows up in fine-grained black mudstones—and looks for all the world as it did when it first occurred millions of years earlier in the same sort of environment, the specimens now preserved 20 miles apart in separate little quarries.

All Hamilton species, brachiopods and clams, snails and trilobites, corals and cephalopods, seem to tell pretty much the same story. That's why a casual collector can identify a brachiopod as *Mucrospirifer mucronatus* in the Marcellus shale and again in the Moscow Formation—two units near the bottom and top, respectively, of the entire Hamilton sequence. Which is not to say that samples of *Mucrospirifer mucronatus* from the two formations are absolutely identical in all detail, that there has been *no* change in the 6 or 7 million intervening years. No one, as far as I know, has studied this particular brachiopod over this interval in sufficient detail to really say that much about its relative stability versus whatever evolutionary change it may have undergone. But what change did occur cannot have been prodigious, for it is as easy and routine for anyone to recognize a *Mucrospirifer mucronatus* wherever it occurs throughout all Hamilton time as it is for any American to identify a bald eagle.

So what's the answer to our patterns of abrupt change out there —a pattern of sudden change, simultaneous throughout the east-central Midwest as far as we can judge from our data? Is the best answer simply to acknowledge the time gap, yet emphasize the stability within both ancestor and descendant, and split the difference, arguing that change seems to come in spurts that are nonetheless *not* overnight? For one thing is clear enough: a reduction from 18 to 17 columns of lenses is a sufficiently slight change that the many thousands (if not hundreds of thousands) of years that slipped by without a record is more than ample for natural selection to go to work in a conventional sort of way and produce our evolutionary change. Convention—with a footnote on the stupendous stability that seems to be the rule—seems to win out.

But clearly that evolutionary change didn't take place right there in the Midwest—for the patently obvious reason that the sea had withdrawn, the habitat had dried up and the creatures, of course, weren't living there any longer. Where had that evolutionary spurt gone on? The much thicker suite of rocks in the East, with its much more complete record of Hamilton time, was a natural place to turn to, and it was there that the real answer, the missing piece of the *Phacops rana* puzzle, turned up. And though the place was predictable, given what we know of the situation in the Midwest, the *time* came as something of a surprise.

THE CONFLICT RESOLVED: A TRANSITIONAL COW PASTURE

The rolling hills along U.S. Route 20, south of the Mohawk River, southwest of Utica, south and east of Syracuse, provide the archetypical exposures of the Hamilton Group. A little quarry, dug into a hillside in a cow pasture and hidden from view along the Morrisville–Peterborough road, sits right near the base of the fossiliferous Hamilton rocks. Around the corner, a roadside cliff exposure shows the very advent of the fauna: grading up from nearly barren rocks, the first signs of open-marine life start popping up—particularly *Tropidoleptus carinatus*, for some reason known to all Hamilton *aficionados* as the true harbinger of normal marine bottom conditions. The term "*Tropidoleptus* fauna" means simply that: an ecological association of species recurring together throughout Hamilton time, wherever and whenever full-blown, normal marine conditions prevailed—in contrast particularly to the stagnant, oxygen-poor bottom conditions that occasionally developed, with their drastically reduced and rather unique faunal elements. When *Tropidoleptus* shows up, most of the rest of the Hamilton brachiopods, clams, snails and trilobites are usually right there with it.

The cow-pasture quarry just a bit higher up in the lower Marcellus Formation is a cut above average in paleontological experience. The clams and snails are particularly abundant—signs, in the Paleozoic, of truly near-shore mud-flat conditions. And the preservation is remarkable: both clam and snail shells typically are aragonitic, and aragonite is a relatively unstable form of calcium carbonate. Calcite is the stable form, and virtually all limestones as old as the Paleozoic have nothing *but* calcite in them. Aragonite simply recrystallizes, given some heat or pressure, as calcite, and one or the other is sure to have gotten to any aragonite in the Paleozoic. When aragonite converts to calcite, crystal size and shape change. The remarkable thing about these calcitic mollusk shells in this particular quarry is that the recrystallization was such a mild affair that the original anatomical fabric of the shells has been retained—at least in sufficient detail that judicious probing with an electron microscope is all that's needed to pick out the details of the delicate original internal structure of the shell.

This little cow-pasture quarry near Morrisville is no trilobite lover's dream. With mostly mollusks there, I spent days collecting with Rollins, kin and students, amassing a hoard of clams and snails, specimens that have served as the basis for several monographs detailing the anatomical ins and outs of these beautifully preserved fossils. But there *are* some trilobites there—*Greenops boothi*, with its star-shaped tail, and particularly *Phacops rana*. Specimens are extremely rare, and usually half a day in the quarry with three or four people poring over the rocks will produce one, perhaps two or three, specimens. But like the rest of the fossils, the few trilobites that do show up are invariably gems, exquisitely preserved. I now have perhaps 50 specimens (at most) from this quarry. And they are the ones that tell us what punctuated equilibria is all about.

One reason, one might suppose, why there are no intermediates between the two midwestern versions of *Phacops rana* is simply an artifact of the way we count. The older specimens have 18 columns, the younger only 17. Maybe it is an either/or proposition, and the possibility of gradational intermediates simply doesn't exist. Were this the case, rapid change from one state to another would seem less surprising and far less of a threat to orthodoxy.

But the cow-pasture specimens put the lie to this supposition: they form one of the very few samples of the *Phacops rana* stock I have ever seen that is variable in terms of numbers of columns of lenses developed in the eyes. There is some evidence that the last column to be added as these phacopids grew up was the one in the very front of the eye. Of the relatively small number of countable eyes recovered from the quarry to date, one has a single lens positioned where the 18th column belongs; another specimen has two (possibly even three) lenses forming part of an 18th column. Yet another trilobite, though damaged, may also suggest development of an 18th column. Further evidence that these early New York specimens are intermediate between *P. milleri* and *P. rana* comes from the multivariate mathematical scores contrasting lens counts with overall head size: *Phacops milleri* specimens typically have many more lenses at any given head size than *rana* samples show—and the cow-pasture specimens (even those with 17 columns of lenses) plot out with the two *milleri* samples rather than with the later *P. rana* samples. The intermediate nature of this sample just happens to span the gap between *P. milleri* and *P. rana* proper—a gap that supposedly records an evolutionary event out there in the Midwest. The only trouble with it is that the intermediate population is a good 3 million years older than that supposed evolutionary "event."

Right above these earliest Hamilton beds, the rocks in central New York have few phacopids—the sandy flagstones produce huge flat clams, some primitive snails and a foot-long trilobite that seems to have loved burrowing around in such coarse-grained near-shore sea bottoms. But there are some *Phacops rana* not too much farther up, and sporadically they occur throughout the rest of the Hamilton—another 7 or 8 million years. In all samples, with very few exceptions, adults have 17 columns of lenses—the standard population number for this species.

That little Morrisville cow-pasture quarry exposes a total of nearly 35 feet of rock. A gray shale near the bottom, the sediments are a bit more quartzy, a bit siltier near the top. Trilobite specimens are available throughout, and the critical question is How much time does this 35 feet of compacted mud and silt represent? Throughout the quarry face, clams and brachiopods occur in what look like their normal life positions, either burrowed into or reclining upon the sea bottom. Such bespeaks a rather calm and gentle accumulation of sediments, not a sudden dumping of thick piles of mud. On the other hand, the mountains shedding all that mud and silt were not very far to the east and kept going up, so we might suppose the streams were constantly charged. But even if we imagine that 1 centimeter a year were deposited—a rather rapid rate of accumulation, but not an utterly unreasonable estimate—we are still talking about 1,000 years for 35 feet to accumulate. That would be the least amount of time for that much sediment to be deposited, and it is just as likely that it took something more like 5,000 or even 10,000 years—not much time in geologic terms, but hardly overnight, either. If what we have here is truly an evolutionarily intermediate population, we are patently not dealing with a case of saltation.

And there is no way to "prove" that we have in fact caught an evolutionary event in midstream, either. But the circumstantial evidence in favor of such a conclusion is overwhelming: only a few quarries sampling the next 6 million years' worth of rock and time produce any variation whatsoever—and none leads to the establishment of a new adult population number of columns of lenses. It seems like a stroke of dumb luck that road material was needed, the farmer was willing and Bud Rollins knew just where the quarry was.

The pattern is clear, and the inference not difficult to draw: that was no evolutionary event in the Midwest. The jump from the *milleri* type, with its 18 columns of lenses, to the newer *rana* type, with only 17, was only apparent. And therein, of course, lies a moral: not all the phenomena of nature are what they may at first seem to be. The sudden transition from "ancestor" to "descendant" was no such thing at all when it seems to take place in the Midwest: The descendant was already up and running around a good 2 or 3 million years earlier. It just didn't live in the Midwest. It never got out there until an ecological calamity—the drying up of the entire inland sea—erased the entrenched inhabitants. Most of the older species made it back when the Centerfield Sea renewed the continental flooding, but the old *milleri* species did not. It never showed up again, anywhere—missing and presumed extinct. What *is* there is a closely related species, what was originally described by the Philadelphia physician Jacob Green in 1835 as *Phacops rana*—a species that, it now becomes clear, had been restricted to the eastern embayments along the uprising Appalachians for the first half of Hamilton time, a species that survived the midwestern marine regression simply because the seas remained intact along the eastern coastline. And it was a species that, once given the opportunity, was certainly able to spread out through the limy bottoms of the coralline seas—as soon as the sea reappeared, and as long (apparently) as the former occupant, the primitive phacopid species with 18 columns of lenses, was no longer there.

This scenario—of two closely related and only slightly different species living in adjacent, but exclusive and nonoverlapping, habitats—relies heavily on the ecological concept of *competitive exclusion:* two closely related and ecologically similar species, with similar adaptations but specialized for slightly different habitats, will divide up a geographic region and exclude one another from their own particular favored environments. It is an appealing image. But it is *only* a scenario, and it is also true that many ecologists these days strongly doubt that competitive exclusion really occurs. Some ecologists even doubt that competition between species ever takes place—a not-too-unreasonable conclusion given the dearth of good examples they can point to in nature.

But there really seems to be a mutual exclusivity going on for 2 or 3 million years: *Phacops rana* (the 17-column form), the newly evolved form, makes a home up and down the eastern marginal seaway, living on the sands, silts and muddy bottoms that typify its chosen habitat. Meanwhile the *Phacops milleri* bunch—the primitive ones, the ones that remained virtually unchanged since the stock's invasion of North America from the Old World province, flourish for those same 2 or 3 million years out on the limy bottoms of the midwestern epicontinental sea. Sure looks like competitive exclusion to me.

And here is a big problem all evolutionary biologists—at least, those who wish to converse across disciplines—face. It is the problem of scale. Ecologists skeptical of the very concept of competition between species, those who claim they simply cannot see any evidence for such raw battling going on nowadays in nature, are fully aware that many closely related species seem to be avoiding one another. Examples like these Middle Devonian trilobites reinforce that impression: species seem to avoid one another, to split up the turf and occupy mutually exclusive subregions, for truly long periods of time. The reason ecologists cannot go out to nature and expect on any given day to witness species actually fighting it out for *Lebensraum* boils down to scale: the evolutionary saga of a species—its birth, its history and its eventual extinction—is played out against a backdrop of hundreds of thousands, often millions, of years. Interspecific competition is an ecological process, an inherently unstable, even intolerable situation which seems to be resolved very quickly—in tens of years, or perhaps hundreds. There are two stable resolutions to such internecine, interspecific warfare. One species will soon drive the other out, in which case the pattern of mutual exclusion is maintained; or there will be an accommodation, a subdivision of resources that will allow members of both species to live in close proximity, if not actually cheek to jowl. At least, that's one answer to the puzzle of why we so rarely see actual competitive bouts in progress in natural circumstances: unstable situations are resolved so quickly relative to the entire time span of the players involved—the two species—that it is inherently improbable that we should stumble upon them in competitive midstream. But we *may* at times, and some ecologists think they have seen some pretty fair evidence for interspecific competition in the wild. For the most part, though, we must rely on the inductive tools of mathematics to simulate the process. And we might ask what other sort of process would lead us to predict these patterns which *look* to us so much as if they reflect mutual competitive exclusion between two species with very similar ecological requirements.

If we follow this line of reasoning a step further, we must conclude that natural selection can act relatively quickly—again, in the context of full-blown geological time. For it is good old natural selection which is the force that must act to resolve competitive situations, be it competition between members of the same species or members of different species. Classic, pure natural selection *is* differential reproductive success among organisms within a single reproductive community, a single species. Some are more "fit," better able to survive and thus more likely to be successful in the game of leaving more progeny. But the tensional situation inherent in competition between two species is resolved in much the same way. Collectively, members of the competitively inferior species will be driven out either physically, or over a period of time: they simply will not be able to maintain a solid breeding population persistently over a number of generations. The competitively superior species will win out because, collectively, its members are more fit, adaptively more able efficiently to exploit the ecological niches for which the two species are in competition. But collective fitness is simply additive, nothing but the average of the fitness of all the organisms within each of the two competing species.

That interspecific competition is resolved by simple natural selection is more obvious when we consider the second possible outcome: accommodation rather than exclusion. When two closely related species are found to be living in the same area in close association, members of each invariably specialize in slightly different foodstuffs, breed in different seasons or otherwise behave in minor but consistently different ways. Of the five common species of the frog genus *Rana* (*Phacops rana* was so named because it reminded Dr. Green of a frog) found near the lakes and streams of the northeastern United States, some, such as the wood and pickerel frogs, typically wander farther from the water than the bull, green and leopard frogs. And though all species may be breeding along the edge of the same pond, their mating calls and exact breeding seasons and behavior all differ substantially. None of the niches of these various frog species overlap to a critical extent, to the point where they are in serious competition with one another. And following conventional theory, it is simply natural selection that produces, maintains and modifies the adaptations of organisms. It is an inference, but a strong one, that it is selection which fine-tunes the adaptive differences between closely related species, allowing them (as it were) to exploit different aspects of a habitat and thus to overlap in the geographic ranges.

Though contentious, the fundamental idea that Darwinian natural selection is the underlying deterministic mechanism responsible for adaptations seems reasonably sound. So does the further inference that competition exists, but is so unstable that its resolution is best thought of as an ecological event—and on the geological scale of things, hardly likely to be too obvious to anyone planning a modern field study. The paleontologist, already inured to far coarser data—in which most of the anatomical details are missing and truly precise timekeeping is impossible—could expect to find even less in terms of hard-core, cut-and-dried examples of unresolved competitive situations. We can hope to see only either/or, before/after situations. Either two species exclude each other or they do not.

Such a situation—in which the data are apparently too coarse to catch the actual event in midstream, at least in the majority of cases—does provide a positive lesson about selection in general, a lesson that should have been obvious long ago. Natural selection—at least, *directional* natural selection, in which a lineage is undergoing concerted genetic changes through time in response to some environmental signal—*simply cannot be the protracted affair Darwin (and many of his successors) thought it must be.* If selection induces a change in state, be it 18 to 17 columns of lenses or, for that matter, larger brain size in our own human lineage, it can happen only if the requisite underlying variation is there, or by chance supplied by new mutations. If these conditions are met, scale enters in again: in a geological sense—that is to say, in "true evolutionary time"—that selective change will ordinarily be *rapid.* But rapid only in the context of millions of years.

And this dovetails nicely with what geneticists have known for half a century: natural selection acts pretty quickly to effect some rather considerable change within a population. When the English sparrow first got to the United States in the 1850s, it quickly differentiated, and today ornithologists recognize a number of distinct subgroups spread out over the United States. The birds differ in relative size of the wings, legs, bill and so forth. As Ian Tattersall and I emphasized in our book *The Myths of Human Evolution,* if an "idea," to put it teleologically, is both desirable and feasible, evolution, through natural selection, will accomplish it, and will do so directly.

The moral is clear: it was a fluke, a pure piece of blind luck, to stumble on that little and hard-won collection of trilobites from the Morrisville cow pasture. It seems to represent a sample of a portion of that *Phacops* lineage which really was in the throes of transition, caught in selective midstream between the stable ancestral species and the descendant species—which also was to become stable and persist unchanged in its eyes and most other features for at least another 6 or 7 million years. And that quarry sample preserves minimally several thousands of years. We have true evolutionary intermediacy here, on a time scale that falls well within the experience of geneticists. (If anything, geneticists are apt to remark how *slow* that transition seems to have been—if my guess is right about how long it took for those 35 feet of sediment to accumulate.) Far from a radical proposition, this transitional population seems just right if we think in the conventional terms of population genetics. This is no example of saltation.

Nor was that Devonian event which was fossilized in Ohio, Michigan and Ontario a genuine *evolutionary* event. The double-barreled and largely contradictory explanations of the apparent jump from primitive to advanced are theories in search of a real event. The interpretation that the 17-column trilobites are the descendants of the 18-column ones still holds, of course, but the real evolutionary event, the actual transition, is sampled in the cow pasture, which records an event 3 million years older than the "gap event" of the Midwest. The latter was an ecological and biogeographic event: one species replaced the other, presumably after the latter's extinction. As it happens, descendant replaced ancestor. But the real evolution, the appearance of the descendant, had taken place millions of years earlier. And the ancestor, in one locale, had simply lived on alongside the descendant, in another, for all those millions of years.

Those two evolutionary explanations—saltation versus gradual transition—after all *do* have a fundamental point in common. Both are theories of the ways and means of transforming the anatomical features of organisms within a lineage—an *entire* lineage. They differ on rate: saltationists see evolving lineages transforming in episodic fashion; gradualists deny true overnight saltations, while admitting that rates may vary within and among lineages. Some lineages change faster than others, and sometimes evolution proceeds more rapidly within a lineage than it does at other times. But the basic expectation, that evolution, at least to a significant degree, involves the wholesale transformation of an entire lineage, is common to both. Though not in itself a totally unreasonable or even unsupported assumption, it does lead to gaffes—such as supposing that the pattern of replacement in the Devonian Midwest was a true evolutionary event. Paleontologists are forever assuming that what they see in one little area, one portion of the range of a fossil species, tells them what was going on in the entire lineage—again, a supposition based on the premise that in evolution, a lineage is invariably transformed simultaneously, throughout its range. What goes on in one place, they think, must be going on all over.

Take the case of the neanderthals, our erstwhile fossil cousins, who have been kicked all over the phylogenetic map since the first discoveries in the mid–nineteenth century. Some paleoanthropologists to this day grumble that neanderthals (and other, more primitive fossil hominids) have been systematically excluded, debarred from the status of being our ancestor, by colleagues unable to bring themselves to seeing such brutes as near kin—let alone our direct ancestors. They call for a restitution of neanderthals to their "proper" place and their recognition for what they are: our immediate forerunners in Europe, the Middle East, indeed all over the Old World. In other words, we need look no farther afield than in the sediments directly below the first signs of modern man (perhaps 90,000 years ago in southern Africa; 50,000 years ago in the Mideast; 35,000 years ago in Europe) for our ancestors. Again, not unreasonable, and the roots of our modern selves, *Homo sapiens sapiens*, surely do lie close to the neanderthals—who are usually classified as merely a different subspecies of *Homo sapiens: Homo sapiens neanderthalensis*.

But in Europe 35,000 years ago, where the archeological record is by far the clearest, what we see is abrupt transition—if transition it be. The skulls and bones of neanderthals, far more robust than the delicate frames of truly modern people, are abruptly replaced by those indistinguishable from our own. And it is not just the bones: the cultural kit bag, the stone implements, changes radically and suddenly as well. No transitions here. Cave art appears, also with no premonitory signs, no early development among the neanderthals. There *could* have been a sudden saltatory leap in body, mind and culture 35,000 years ago in Europe from the neanderthals into ourselves, but the simpler explanation surely is invasion and replacement. The transition may have occurred earlier in the Mideast, where two caves in Israel have actually produced what look like intermediates. Or perhaps the transition between early neanderthal-like hominids and modern people actually occurred still earlier in southern Africa. It is not revulsion against brutish ancestry that leads one to reject the saltatory origin of modern man in Europe 35,000 years ago.

In fact, if evolutionary biology has established one solid fact over the past 100 years, it must be that evolution within species does not typically proceed in lockstep fashion, everywhere affecting all members within a far-flung species in the same way at the same time. Species, rather, differentiate, and the process of speciation itself is a fragmentation of one species into two or more. There is every reason *not* to expect the wholesale, full-scale transformation of an entire species. And all apparent cases of sudden transformation in the fossil record should immediately be suspect. In paleoanthropology, the view that evolution must be the linear transformation of one entire species into another reached its peak in the writings of

Carlton Coon, who, in trying to understand the data and yet keep the idea of linear adaptive transformation intact, was forced to conclude that *Homo sapiens sapiens*, modern man, evolved at different rates in different parts of the world, appearing first in China, then in Europe and finally in Africa. But this was not, in Coon's view, the evolution of *Homo sapiens sapiens* in one place (say, China) and its subsequent spread: it was a direct transformation of the human lineage that left the more primitive, neanderthalish stage at different times as the pace of evolution differed from place to place. What this says is that modern people evolved not once but three or four times—a fantasy even less in line with biological common sense than the simpler notion that evolution slowly modifies an entire species as time goes by. As we shall see, species have definitive evolutionary histories, including beginnings, life spans and (inevitably) terminations; they are not simply end points on some sliding gradient of transformation. And at the very least, we must carefully distinguish between ecological events—in which one species, even if it is the descendant, simply replaces another—and true evolutionary events, which in the very nature of things will be far harder to spot in the fossil record.

ONCE MORE UNTO THE BREACH

History does tend to repeat itself, and recurrent patterns give us the data we seek to explain in science. The evolutionary history of the *Phacops rana* group was not complete with the reduction from 18 to 17 columns of lenses. Late in the 8-million-year interval of the Middle Devonian, a rather similar event happened once again. In Michigan, a trilobite that E. C. Stumm called *Phacops rana norwoodensis* (it is actually a distinct species, *P. norwoodensis*) shows up in the Petoskey Formation. This trilobite—definitely a member of the *P. rana* stock—has but 15 columns of lenses. Back in New York, in the Tully limestone, there is again evidence of variation—this time between the ancestral population number (17) and the more common 15 column specimens. Some trilobites have 16. For example, most specimens from a quarry on Portland Point on the east shore of Cayuga Lake have 15 columns—but two specimens have the ancestral complement of 17. Specimens from other localities in New York occasionally show either 16 or 17, rather than 15, columns. Once again the evolutionary action seems to have been concentrated in the East and not reflected in a sudden anatomical leap in the Midwest, a saltation down from 17 to 15 columns.

But the example lacks some of the compelling qualities of the older event. It is impossible to show with certainty, for example, that the variable population of the Tully limestone near Ithaca, New York, is older than the apparently fully evolved sample from Michigan (or, for that matter, the samples from Iowa and Wisconsin). The same sort of event seems to be going on, but it is less certain that evolution occurred in the East, ancestor and descendant living on side by side for a time until the descendant was finally able to "go west," replacing (or displacing) its ancestor. That seems to be the case, but the time resolution is simply not as clear-cut as in the earlier event—the evolution of *Phacops rana* from *P. milleri*, and their subsequent interactions.

SUMMARY: AN EVOLUTIONARY PATTERN

So, the bare bones of the case of *Phacops rana*, its ancestors and descendants, are simple enough. Sometime around 380 million years ago, as Europe, Africa and North America came together, a number of new invertebrate faunal elements—in most cases snails, clams, brachiopods, corals and trilobites well established in the Old d province—came into North America. A vast sea spread over the continental interior for the first time in perhaps 30 million years. *Phacops milleri*—nearly identical to the slightly older *P. africanus*—spread throughout the early Hamilton sea. The primitive 18-column-form *milleri* alternated with its variant form *(crassituberculata)* in slightly different habitats of the Midwest, lasting for some 3 or 4 million years. But its existence in the eastern, near-shore muddy- and silty-bottomed habitats was far briefer: the very earliest sample so far found reveals an intermediate population that seems to bridge the gap in terms of number of columns of lenses and in other features between the ancestral form and the descendant that was to persist for so long. Once the number of columns stabilized at 17, this descendant form—*P. rana* proper—quickly spread down the eastern marginal sea and remained there (in what is now central New York State on down through the southern Appalachian states) for the rest of Hamilton time, approximately 7 million years.

The picture of stability and "peaceful" coexistence of the two species in adjacent habitats persisted for about 3 million years, until the sea withdrew from the continental interior. *Phacops milleri* and all other inhabitants of the inland sea, of course, disappeared with the sea. When the sea returned in the widespread "Centerfield" inundation, *P. rana*, which had been living in the eastern marginal sea, came out with it—further inferential support for the thesis that it *could* have lived out there all along but for the competition posed by presence of its close (ancestral) relative *P. milleri*. In any case, for the next 3 million–odd years, *Phacops rana* remains the sole representative of its stock, sole *rana*-esque component of the entire Hamilton sea. Then there was another event, though its details are less well worked out: *Phacops norwoodensis* appears in the Midwest, the apparent offspring of *P. rana*. Again, a variable sample was collected in the East, once again in central New York—a variable sample that once again bridges the entire (if modest) amount of evolutionary transformation.

Thus the pattern of change is different in the two regions. In the Midwest, three species follow in order up the geological column: *Phacops milleri*, followed by *P. rana*, followed by *P. norwoodensis*. They do not overlap in time there. Each was stable throughout its temporal range—especially in the very characteristics (mainly, but not exclusively, the eyes) that tell them apart in the first place.

The story in the eastern marginal seaway was different. Twice, variable populations link up a descendant species with its ancestor. The transitions seem to take several thousands of years, and were definitely *not* overnight, saltational phenomena. But it happened, at least in the first instance, millions of years before the pseudoevents—with their alternative interpretations—in the Midwest.

So much for the bare bones of the case. Calling them "facts" would be gilding the lily, but 15 years of rechecking has failed to budge the outlines of the story. The case hinges on the correct interpretation of the *ages* of all the samples (interpretations based on other paleontologists' work on different fossils); also critical is the correct interpretation of the evolutionary relationships between the various samples (an analytic task greatly aided by the stability, the utter sameness, that easily links up most samples into a few discrete species). The scenario depends, too, upon the validity of the supposition that those two variable populations really have caught, however accidentally, an actual evolutionary event in midstream. The pattern seems clear, and this basic interpretation eminently reasonable. But it could be wrong—though continued collecting over the past 15 years has only buttressed the pattern that first became evident in the late 1960s.

The rest is even more inferential. Why did that lens reduction occur? I have always *assumed* it was an adaptive change, the product of natural selection. When forced to (usually when questioned after a lecture), I even have a ready scenario for the adaptive significance of that lens-column reduction, but it is a scenario that remains very much untested: I think the first column of lenses became obsolete as the lateral margins of the head—the glabella, which housed the stomach of these creatures—expanded a bit. I think some of the trilobites started looking at their own heads, and the removal of that anterior column became "desirable" if not an absolute necessity. I confess I do not really know. But it is important to emphasize that the time it took for that change—neither a one-generation leap nor an arduous process over millions of years—seems just about right for natural selection. This little story of *Phacops rana,* its ancestors and its descendants requires no special theory at all to explain it—once we relax the mythic expectation that evolution always need be the wholesale transformation of an entire stock, an inevitability given all that time to play with. There is a far simpler explanation for this pattern of episodic changes and shifting distributions—an explanation that comes from a different corner of evolutionary biology.

4 THE EMERGENCE OF PUNCTUATED EQUILIBRIA: SPECIATION AND EVOLUTIONARY CHANGE IN ANCIENT SEAS

There is an old saying "The present is the key to the past." Charles Lyell, who founded the modern science of geology with the publication of his three-volume *Principles of Geology* (1830–33), enjoined his colleagues to use their heads and common sense: we can understand much of the earth's past if we simply assume that the processes we see acting around us today were also operating in the remote past.* Thus we see stream beds laden with silt and clay, and a little investigation shows us that the particles come from soils and, ultimately, bedrock somewhere upstream. Water runs downhill (gravity at work) and stream banks erode to add to the suspended sediment load, but wherever the stream slows, as in the inner side of a bend in a sinuous course, or where it reaches a lake, the heavier particles drop out and collect on the bottom. Eventually the hills are worn away and the basins fill up. If sea level remains constant and the lands are not uplifted in a renewed period of mountain building,

* This is the doctrine of uniformitarianism—the notion that natural laws operating today were at work in the past and will continue to hold in the future. As S. J. Gould first pointed out (in 1965), this aspect of "uniformitarianism" is not unique to geology: *all* science assumes the constancy of natural law. The other side of uniformitarianism, the unwarranted assertion that all geological (and biological) change takes place at a characteristically slow and constant rate, is demonstrably false. Few processes other than radioactive decay are utterly constant, and even the pace of the erosion of gullies into canyons is in part episodic. "Punctuated equilibria," of course, expressly considers the episodic pace of evolutionary change that we see in the fossil record, and seeks its explanation in biological processes known to be operating today—"uniformitarianism" in that sense, but obviously not in the sense of "uniformity of rates."

a whole lot of geologic history can be deduced from rocks on the simple assumption that these simple activities have always operated on earth. After all, they require only gravity and the presence of water and land on the primordial earth. And the very existence of sedimentary rocks that are more than 3.5 billion years old buttresses our faith in such simple assumptions.

If paleontologists have made one general sort of mistake in contemplating evolution since Darwin, it has been the tendency of some of the more adventurous thinkers among them to invent novel explanations of *how* evolution takes place. Starting, frequently, with genuine difficulties in reconciling natural selection with their perception of change in their fossil bones and shells, some paleontological theories dreamed up to replace orthodox Darwinism have shrugged off the normal constraints of real-world experience; an active imagination can contrive any number of evolutionary mechanisms. But if we have no way of testing them, of seeing if some process or other really does occur in nature, we are doing an end run around one of the cardinal rules of science: we *must* be able to bounce our ideas off firm evidence—or otherwise anything goes.

But this is a tremendously tricky proposition, this business of testability in the study of evolutionary processes. When S. J. Gould and I, in 1972, wrote the paper that coined the term "punctuated equilibria" and set out its basic characteristics—the patterns of stasis and episodic change, and the biological theory we thought best explained the total pattern—we said that "no theory of evolutionary mechanisms can be generated directly from paleontological data." This was one of the more infuriating statements in a paper that in general annoyed a lot of paleontologists, for it seemed, once again, to remove paleontology from the direct ranks of sciences concerned with establishing just how it is that life actually evolves. There is a delicate balance here, though: we were simply trying to avoid the not-unreasonable relegation to the lunatic fringe that some paleontologists in the past had suffered when they too saw something out of kilter between contemporary evolutionary theory, on the one hand, and patterns of change in the fossil record on the other. Gould and I wanted our ideas to square somehow with what biology in general thought it had come to understand—by the early 1970s—of the nature of the evolutionary process. We simply felt that knowledge *is* cumulative; and the best way to win friends and influence people is *not* by setting one's discipline apart from the rest of biology, insisting that an utterly new set of rules must be applied—that somehow biology had it all wrong and that we need a completely new and different theory.

Yet our paleontological brethren—S. M. Stanley was one of our earliest critics on this very point—were also right: you *can* test notions of process against the fossil record, and it is possible for paleontologists to come up with ideas of how evolutionary mechanisms operate, ideas that would yield predictions of results, patterns of change, that could be checked against the fossil record. The problem boils down to this: expressly paleontological theories of change must yield specific predictions of the sorts of things one can reasonably expect to experience with fossils—patterns of stability and change in hard-tissue anatomy of plants and animals over broad areas and truly formidably long chunks of geologic time. All else is out-of-bounds: theories of large-scale mutation, for example. It is an utter irony that "punctuated equilibria" has been confused with the ideas of Richard Goldschmidt on "macroevolution" (see Chapter 3) when by the very rules of the game paleontologists *cannot* say anything whatever about the rules of genetic transmission or genetic change. Our data are simply too coarse, too gross for us to "do" genetics with fossils. We are constrained to take genetics—theories and data from geneticists—as given. And we are constrained to operate within the bounds laid down by geneticists. There is one single "evolutionary process," and whatever it is geneticists appear to have established as generally "true" of the evolutionary process cannot be violated by any additional, or new, theory coming from paleontological quarters.

And therein lies the mistake made by some earlier paleontologists who felt no compunction whatsoever about quarreling with genetics. When Henry Fairfield Osborn, for example, was writing in the early decades of this century, genetics was still relatively young and something of a presumptuous upstart. Both president and director of the American Museum of Natural History, Osborn was a powerful and productive figure in early-twentieth-century paleontology. He was also wealthy and prone to a snobbish elitism that came out in some of his evolutionary views. Rather than seeing the rich and influential as the winners of life's competitive struggle ("survival of the fittest" has, after all, served as a catchall rationale for many a robber-baron entrepreneur), Osborn, whose wealth was inherited, posited an *internal* mechanism, an inherent drive toward improvement. Osborn wrote of "aristogenes"—genes that perhaps produce aristocrats among men and simply superior organisms among animals and presumably plants. Osborn saw evolutionary change as governed by direct modifications of the germ plasm—his "aristogenes." Geneticists, needless to say, had a hard time swallowing *that* particular concept. On the other hand, some of Osborn's difficulties with prevailing evolutionary thinking in the 1920s were based on a dissatisfaction with natural selection as a blanket explanation for absolutely everything that has happened in evolution.

And some of Osborn's problems derived from an even more common and vexatious source: basic misunderstandings between two disciplines. Both genetics and paleontology obviously bear on evolution, but otherwise there are no two more dissimilar activities than the finding, collecting and analysis of fossil bones and shells on the one hand, and figuring out the physical and chemical basis of heredity on the other. To this day the very elements of each discipline remain a closed book to the majority of practitioners of the other.

Theodosius Dobzhansky, a geneticist and the closest thing to the "founding father" of the "synthetic theory of evolution," persistently criticized (derided is actually more like it) Osborn for writing: "Speciation is a normal and continuous process; it governs the greater part of the origin of species; it is apparently always adaptive. Mutation is an abnormal and irregular mode of origin, which while not infrequently occurring in nature is not essentially an adaptive process; it is, rather, a disturbance of the regular course of speciation." It took another paleontologist, George Gaylord Simpson, to point out that Osborn was simply using the word "mutation" in its *original* sense. The word first entered the paleontological literature in a paper by W. Waagen on ammonite evolution—a paper written in 1868. Waagen was discussing a series of ammonites (extinct molluscan relatives of the modern pearly nautilus) that he felt formed a linear ancestral–descendant lineage. Successive geological strata yielded slightly different forms, and these Waagen called "mutations." To "mutate" is simply to "change." "Mutation" simply popped up again as the natural choice of a word to describe spontaneous modifications in some anatomical feature of an individual organism—thus, presumably, a change in the underlying genetic material—when geneticists started plumbing the depths of heredity in the early 1900s. Simpson was setting the record straight: neither

Osborn nor Dobzhansky was aware of how the other was using the word "mutation." To Dobzhansky, mutations were the ultimate source of all variation. Following Waagen's usage, Osborn saw mutations as *changed forms,* and as such the results of the evolutionary process as much as a part of that process itself. And their suddenness, conflicting with the gradual, progressive change one expects under natural selection, led Osborn to characterize mutations as "abnormal," "irregular" and "not essentially an adaptive process." Dobzhansky saw mutations as the first essential step in the process of adaptive change: mutations are the ultimate source of the genetic variation on which natural selection works.

But ignorance of the law is no excuse. In 1944, Simpson published his *Tempo and Mode in Evolution*—with the express purpose in mind of effecting a synthesis of genetics and paleontology. Simpson's logic was impeccable: there is but one "evolutionary process," just as there has been but one evolutionary history of life on earth. What genetics reveals about how the process works must fit in with what paleontology reveals about what that history was like. There can be no inconsistency—and Simpson set out to show that what was known about the fossil record was consistent with what was known about genetics in the early 1940s. He also argued—forcefully—that no additional vitalistic, internal, mystical forces need be entertained as evolutionary mechanisms. But what Simpson did insist upon was that the peculiar patterns of evolutionary change which could be seen only over the truly long spans of time in which evolution worked—and that means looking at the fossil record—demanded theories of evolutionary processes which used the standard ingredients of genetics (mutation, natural selection and other factors) but in combinations and in degrees undreamed of by geneticists who were handicapped by the small-scale changes they could study over the briefest of time intervals. Simpson brought paleontology back into the mainstream of evolutionary biology—all the while insisting that paleontological phenomena had much to tell geneticists about the true nature of the evolutionary process.

Simpson's lead seems logical—and seemed the best course to follow when alternative explanations were considered for stasis and episodic evolutionary change in our trilobites, clams and snails. But there was, and still is, a nagging doubt: are *all* the processes of evolution, those which promote stability as well as those which cause change, to be found in the DNA of organisms and the shuffling of gene frequencies in populations? Can the fossil record go one step further to suggest that additional (not *alternative*, but *additional*) processes are also at work? Our answer in 1972 was "Well, yes and no." The basics of the punctuated pattern are easily explained by a segment of evolutionary theory that had not been extensively applied to the fossil record before. We simply drew on a different set of well-established evolutionary principles from the ones conventionally used (by zoologists as well as paleontologists) in thinking about fossils in evolutionary terms. We invoked *speciation*—the formation of two species from a single ancestral species, a process that had been thoroughly analyzed and was well ensconced in the literature of evolutionary biology, largely through the efforts of Ernst Mayr. Speciation involves more than simple, linear adaptive transformation of genes, behaviors, physiologies and physiognomies. It is, quintessentially, a sundering of one coherent, integrated reproductive community into two (or, occasionally, more) discrete daughter reproductive communities. Adaptive change is usually implicated, but adaptive change alone is not sufficient to form two reproductively discrete species. And it is no exaggeration to say that all (save the metaphysical or otherwise arcane) paleontological models of evolutionary change were couched strictly in the language of the adaptive transformation of anatomies through time—with natural selection as the prime mover.

But in taking the simple step of applying the theoretical model of geographic speciation to explain our paleontological data, we unwittingly stumbled onto a paradox: we had gainsaid what had seemed a perfectly acceptable explanation of long-term trends, seemingly directional changes that go on for tens of millions of years, affecting most, even all, of the species within a lineage. The fossil record really does seem to be hinting at the need for some additional theory—a kind of process so coarse that it could be detected only through the fossil record. It is this latter implication of punctuated equilibria which remains controversial today with evolutionary biology. But back in 1972, merely interpreting the fossil record in terms of speciation was enough to provoke howls of outrage among some of our paleontological brethren.

SPECIES AND SPECIATION

We sometimes have trouble seeing things that are bigger than we are, things that are vast and perhaps last for millennia as definite entities. It's all a matter of scale, of course. The galaxy Andromeda, our nearest neighbor beyond our own Milky Way galaxy, looks coherent enough, if a bit diffuse, in a home telescope. Its vastness is comfortably and comprehensibly shrunk by the magnitude of its distance from us. But here on earth it is tough to turn that telescope around, reducing huge things and letting us see them for what they really are. Large objects are more intractable than smaller ones. Electron microscopes show us the fine structures of the endoplasmic reticulum of our cells, the factories where proteins are assembled. We can now photograph atoms, and bubble chambers record the evanescent spurts of elementary, subatomic particles. But we have no machines that focus on items as huge as species. We probably know more about the general nature of atoms than we do about species. Indeed, at any moment, seemingly at least half the world's evolutionary biologists are perfectly prepared to deny that species —*any species*—even exist.

And, of course, that was essentially Darwin's position: we speak of "species" as a handy way of talking about the elementary divisions of organic nature, but hardly anything more. Species were "fixed" to the pre-Darwinian naturalists. Species fixity went hand in glove with divine creation of all separate "kinds" of life (indeed, "species" is the Latin word for "kind"). Darwin ended all that, and to him and many later biologists, the notion of evolution blew not only "species fixity" out of the water, but the very idea that species even exist in any meaningful sense. And we have already seen how the very notion of evolution has led biologists as well as paleontologists to expect change to be so inevitable, so much a necessary aspect of the natural world, that once a new species does appear it will soon become almost unrecognizably modified. To convince the world that life had evolved, Darwin had to counter the erroneous notion of "species fixity," which Darwin realized was the very antithesis to the general concept of evolution. But in his zeal to overcome that hoary notion, Darwin painted such a picture of fluidity that species no longer seemed to exist in any special or specific meaningful way.

What are species, anyway? We need something of a *Gestalt* switch here, because one's view of what species are very much depends upon how one looks upon them. My switch hit me in, of all places, the football stadium at Berkeley as I watched the Class of 1965 graduate. A fly landed on my right knee—an ordinary housefly. *Musca domestica,* I absently mused (I was just getting used to slapping Latin binomials on animals and plants and realizing that most creatures so far encountered by man had such a moniker). But then I began to think, and an entire set of principles I had learned but never consciously applied to the "real" world came flooding in. I started thinking about all those other flies out there—houseflies just like this one. Or near enough that I wouldn't be able to tell them apart without looking very closely. Flies that all look pretty much the same, and are consistently different from all other flies. Flies that hatched from the same cluster of eggs, or from nearby clusters laid by females who *did* come from the same cluster of eggs. I began to think about that reproductive plexus, that complex of mating and egg laying which tied all those flies together. That lone little Berkeley fly came to symbolize a vast legion of very similar and all at least vaguely interrelated flies. I began to get a feeling—and it was little more than that—for what a species is: *Musca domestica* is a specific name, but for the first time for me such a name stood for something more than a mere abstraction of a particular set of anatomical properties shared by a few specimens. It stood, as well, for an entire community, with perhaps millions, even billions, of members at any one moment. And those members were united not only because all of them had those unique anatomies and behaviors setting them apart from all other "kinds" of fly. They also belonged to a far-flung reproductive network. They all, by hook or crook, were involved in spreading genes around within their community—genes destined never to be further shared with any other species save some possible and as yet unborn descendant species.

Think of a species. It is like sending Raquel Welch back in her cinematic *Fantastic Voyage* into a human's bloodstream. All those phantasmagorical red blood cells, rapacious white blood cells—all running around disconnected, rushing singly and severally down the pike. But one thing is certain: all have a definite shape. Each blood cell has been blown up to the point where its anatomy is obvious. When a major popular science magazine recently ran a scanning-electron-microscopic photograph of a red blood cell, a diatom (a silicious, one-celled alga) was seen stuck in its center—an image blood experts frankly had a hard time puzzling out, until someone pointed out that the specimen was oddly and inexplicably contaminated. All objects, big or small, when brought into human dimensions and thus into focus, take on some sort of interpretable shape.

Another thing Ms. Welch encountered was space: big as those cells and antibodies were, there was far more clear plasma than cellular bodies in that bloodstream. Once the blood was magnified and no longer seemed to be "just liquid" the way our ordinary senses proclaim it to be, once the individual bodies that make up the solid matter of blood had appeared, the next obvious thing was the *space* between the particles. It's the same, apparently, with just about anything. Viewed at the proper scale, most things turn out to be mostly space—the universe being the most renowned example. But the internal structure of galaxies, or our solar system, follows the rule as well. That's all well enough—at the level of the solar system, at least, it is obvious that matter is condensed into a sun, nine major planets (at last count), their satellites, an asteroid belt and comets. The rest—the vast rest—is empty space. Or consider an atom, say an atom of hydrogen, with one electron whizzing around a proton—at some considerable distance given the size of the particles involved. Protons, it turns out, are mostly space as well. Most of the universe, most of the discrete *entities* in the universe, is space. Sometimes it's obvious: the Adirondack forest I can see as I write this is mostly space, as thickly and densely set as those trees appear to be. Failing to "see" species is precisely like missing the forest for all the trees.

One reason, I think, why biologists persistently and consistently refuse to see species as entities is that we are all in the position of Ms. Welch in the bloodstream. We are tiny, and now we are looking at some monstrously huge structure, like a kidney, and trying to figure out what on earth it might be. We see the individual tubules, and perhaps even that they bear some relation to one another, banding together to form some large structure whose outlines we can only dimly perceive and whose function we can only surmise.

The problem we have in seeing species partly devolves from considerations of vastly different scale between the extent in time and space typical for humans, on the one hand, and species, any species, on the other. And it also comes from our very perceptions of what we ourselves are, every one of us: individual organisms. We sense ourselves as integrated wholes. It is only by an effort of will that we can see ourselves as composed of parts, or organs, tissues and cells, or proteins and other sorts of chemical substances. That we are mostly water is a long-standing joke. That we are mostly space is less well appreciated. But it is both necessary and totally appropriate for present considerations that we *do* see ourselves as coherent entities. The boundaries of a human body (or that of any organism save a very few) are sharp, abrupt, discrete. Our skin is our wrapper, our interface between the external milieu and our inner selves. We are, without question, definitively bounded in space.

Then too, organisms are invariably born—or otherwise spring into being, as when a simple parental hydroid living in a lake buds off a miniature version of itself by asexual fission. No organism lives forever, and that means not only that all eventually die, but that all are born as well. The span of time between birth and death is finite. This temporal boundedness of organisms is as clear to all of us as the spatial boundedness provided by our skins.

All of this "boundedness" of organisms, obvious as it is, is an important aspect of the singularity of an organism, the properties that help make it an "individual." Recently biologist Michael Ghiselin (1974) and philosopher David Hull (1976) have formally proposed that species can be construed every bit as much as "individuals" as can organisms. They ask us to see species as historical entities with definite boundaries. As I have spelled out in my *Unfinished Synthesis*, the implications of this notion for evolutionary theory in general are profound. For the moment, we must simply ask how it is that we can think of species as "individuals," when clearly any given species is composed of *many* separate (usually) individual organisms.

And in many ways the analogy between organisms and species seems farfetched. Where, for instance, is the analogue of the skin that might give a species its spatial boundedness? There is none, at least nothing that can serve as a direct analogue to the skin on our bodies. Yet most wildlife field guides sport maps showing the distributions of at least some of the species included. And these maps have definite dark lines circumscribing the known limits of distribution of these species. More detailed localized maps are more precise. And range extensions and contractions are continually going on: armadillos and mockingbirds have been moving north, beyond their "usual" southern habitats, for several decades now. Coyotes have been moving east and becoming more numerous, while moose are once again establishing a toehold in New York State—where they had been extirpated in the 1880s (and where their extirpation, in true bureaucratic Catch-22 fashion, prevents their being included in the State's current endangered-species list!). There is no "skin" —the outer limits of a species' spatial distribution occur at the outermost reaches colonized by some of its component organisms. But plotted on a map, typical species distributions are *coherent:* that is, one can draw an irregular boundary around the spatial distribution of a species, beyond which it is not known to occur, and *within* which it occurs everywhere that its favorable habitat is developed.

No species occurs everywhere on earth. Not even *Homo sapiens,* our own species. Our cultural accoutrements have enabled us to conquer and inhabit a truly vast range of terrestrial environments, though those on the polar ice fields require special lifelines of support. Our invasion of the oceanic and atmospheric realms is tenuous and always temporary, and likely to remain so. The rest of the biota must still rely on their own physiologies, anatomies and behaviors to cope with the physical world (and with other organisms—the two together are the "environment"). Ordinarily this means that organisms are restricted to a particular geographic area by a combination of physical barriers, resource restrictions and physiological requirements.

Some species are quite widespread. Moose, species *Alces alces,* the largest living member of the deer family, despite their troubles getting back into New York, occur throughout northern North America *and* Europe. Wolves, *Canis lupus,* are likewise circum–northern hemisphere. Why, then, do they not occur entirely throughout these northern-hemisphere continents, and indeed, in South America, southern Asia and Africa as well? Much of the answer lies in a species' history, what happens to it within the course of its life span. Many species simply remain where they are for as long as they exist because the niches to which they are adapted remain stable and in the same basic area. Even if a species could find a suitable habitat elsewhere, it might well not be able to get there. Closer to home, though, and in a more immediate time frame, there is also the practical matter of limitation based simply on an organism's needs and tolerances. Temperature is an obvious environmental factor that has much to do with limiting the distribution of a species. Water availability is another. Desert organisms—insects, plants, reptiles and mammals—all face especially stringent conditions of water supply and extremes of temperature ranges. All desert dwellers, for example, restrict the amount of water vapor they exhale, though only ostriches so far have been shown to manage 0-percent humidity in their exhalation. The adaptations for such environmental extremes simply make desert dwellers ill suited to life near a swamp, or in the arctic, or anywhere else that the special conditions to which they are adapted do not exist. The converse, of course, is equally true: lack of the special physiology required for desert life precludes any organism from taking up residence there.

Deserts are an extreme, hence obvious example. The fine-tuning of adaptations to meet the basic demands posed by an environment are legion and often subtle, but of profound and material significance to the distribution of organisms—and so to the distribution of species. The same is true of energy resources. Plants, which photosynthesize to make their own food, require a certain mix of nutrients, sunlight and water—and different plants require different mixes. Animals eat plants or other animals, and the availability of one's food preferences very much determines where one can or will go—a generalization not excluding humans. Adaptation, in other words, is as much concerned with the procurement of energy as it is with coping with the physical exigencies of the environment. Adapted as all organisms are to a certain subset of all possible energy sources, their distributions will naturally coincide with the limits of distribution of their food resources—and, in the case of animals, the food "resources" are other sorts of organisms that are *themselves* limited in the very same way.

Physical barriers defining the outer edges of a species' distribution are perhaps the easiest to see. Some are striking, such as the landlocked status of the enormous tortoises that Darwin encountered in the Galápagos Islands on his *Beagle* voyage in the 1830s. Each species is absolutely stuck on one of the islands; it is easy to discern on which island a specimen was collected—a clue, indeed, to the way in which the evolutionary process works.

In the seas, organisms tend to be restricted to huge water masses, bounded where two conflicting current regimes come together. Distinct species of the microscopic, unicellular protozoans and algae that float near the surface of the world's oceans, for example, frequent the mid-latitudes, distinct from closely related, but demonstrably different, species just south of the "Antarctic Convergence"—the boundary between the polar waters and those of the southern Pacific. And on land, too, many organisms, even entire species, are confined by mountain ranges, canyon walls and rivers, if they lack the wherewithal to jump, swim or fly over and around them.

And then there is competition. As we have seen in the last chapter (3), some biologists even doubt that species compete with one another—though they readily acknowledge that organisms *within* a species can and often do compete with one another. Competition for resources and *Lebensraum* within a species, of course, is the very stuff of natural selection. Competition among members of different species, it is said, for precisely the same sorts of commodities can lead to definite limitations of the distributions of different species—as I was speculating for the various species of *Phacops* back in the Middle Devonian sea. Whether such a dynamic leads to selection on a "species scale" or not remains a thorny issue in contemporary biology—an issue I'll take up in Chapter 5.

Well, what about species' "spatial boundedness"? Species lack skins, yet at any one moment, at least, they appear tolerably bounded. The boundaries are at best flexible, so perhaps the best analogy after all is with an individual amoeba, not an individual moose: amoebae are forever creeping along, modifying their external shape as they send their pseudopodia (literally "fake legs") out this way and that to search out and engulf food. Species distributions are not *that* plastic, but species do send out "lobes" here and retreat there just like an amoeba. The point is that species are spatially bounded at any one moment, even without having anything remotely like a "skin." And the *Phacops rana* story suggests that the spatial boundaries of a species—the outer limits of its distribution—may, and in fact usually do, change through time. But changeable though the boundaries may be, a species always remains so bounded.

Pursuing the analogy a bit further, we must now look at the temporal boundedness of species. Are they as discrete in time as they are in space, enhancing their candidacy as a sort of individual? In other words, do species have births and deaths? The quick answer is a resounding "Sure they do": if the fossil record has anything at all to tell us about the history of life, it is that the species of 600 million years ago, or 400 million, 200 and 100 million years ago, are all different from those we have on earth today. By 1 million years ago, as we approach the present, things are a bit different: most of the marine clams and snails from the Purisima Formation in California, remember, are alive today, and the Purisima is several millions of years old. Yet our own species, *Homo sapiens*, had not yet appeared as the Purisima sands were accumulating. We can conclude from such observations that most species living in the past, probably including all those which were alive more than about 10 million years ago, are extinct. Since life is "alive and well" and living on this planet, and since we cannot find modern species in rocks more than 10 million years old (which, I suppose, *could* be construed as negative evidence), we must further conclude that species undergo a "birthing" process as well as a "death"—or extinction—process. Beyond those simple conclusions, the problem gets tricky.

Extinction, on the whole, is more straightforward than the origins of species. It is far more obvious, for one thing. The past 570 million years of the history of life is fraught with entire episodes of extinction—not just the odd species dropping out, but entire lineages coming to what looks like a fairly abrupt halt. When creationists, those who would substitute a literal reading of Genesis for conventional "evolution science" in the science curricula of our secondary schools, tell us that the "geologic column" is an artifice, a fiction geologists have concocted to establish and defend the very notion of evolution, they blithely ignore the fact that the earliest geologists were for the most part themselves creationists. Most pre-Darwinian European scientists were creationists. The early geologists detected a gross sequence in the occurrence of fossils in their rocks. To an amazing extent, the sequence seemed the same virtually the world over. What we have in the fossil record is a series of more or less devastating disruptions of the world's ecosystems. The less pervasive, those extinctions which affected only some regions, necessarily have left a less dramatic story of annihilation than some of the truly major extinction events, in which scarcely a single species made it across the boundary that is now drawn by the nearly simultaneous disappearances of so many species. For that is one practical advantage that extinction, however grim it may seem, has given us: the broader, more pervasive an extinction, the more species that have succumbed, the more nearly worldwide the extent of the calamity, the more marked the division between "life before" and "life after." The early geologists simply recorded what they saw, so that John Phillips was able to specify the Paleozoic, Mesozoic and Cenozoic eras as the three major empirically based subdivisions of time since complex life had first appeared on earth. Phillips coined the names in 1840, nineteen years before Darwin published the *Origin*. (Adam Sedgwick had used the term "Palaeozoic" before Phillips named the geological eras.) He had no idea that his Paleozoic ("ancient life") began as long ago as 570 million years. Lesser divisions of geologic time simply mark episodes in which extinction has been of lesser magnitude.

D. M. Raup has recently estimated that perhaps as many as 90 percent of all the species then on the face of the earth vanished in the great extinction at the end of what we call the Permian Period—the last division of the Paleozoic Era. This great Permian extinction was probably the biggest ever (as yet) to hit the biota—and the 90 percent of, say, 5 million species means that about 4½ million *species*—not organisms—died out.

There are several arresting features to these mass extinctions. Luis and Walter Alvarez, father and son, have documented the high level of the element iridium at the Cretaceous/Tertiary boundary, where rocks of very latest Cretaceous age are overlain by hardened sediments of the lowest Tertiary. The events at this boundary amount to history's second-biggest extinction. The Alvarezes suggested that such a gross concentration of iridium, far out of proportion to its normal occurrence on earth, points to an extraterrestrial origin of the layer. Indeed, meteorites are sometimes known to harbor such disproportionately high quantities of iridium. The Alvarezes conjectured that a huge asteroid had struck the earth, probably somewhere in the sea, vaporized and sent up a huge cloud of gas and dust that occluded the sun, stymied photosynthesis and thus triggered the ecological collapse that *is* a major extinction. They posited an ecological disaster in *ecological* time—10 years, say, to wipe out most of the inhabitants of the globe. But the fossils seem to suggest that extinctions of such stupendous magnitude, degradations of virtually all the world's ecosystems, take longer than 10 years—something on the order of a million years or even more. And few comparable iridium layers have been found associated with other mass extinctions. Such catastrophes, it would seem, take more than a scant 10 years—though they look sudden in the vaster context of the hundreds of millions of years of life's history.

And extinctions cut across lines of genealogical affinity. Extinctions are truly and thoroughly ecological because the species that drop out are insects and dinosaurs, grasses and palms, ammonites, fishes and planktonic microorganisms awash in the sea. So many species go that *all* the species of a certain genus—say the genus *Tyrannosaurus* or of the genus *Triceratops*—may perish. All the species of a series of related genera may go, as did all the species of the genera *Scaphites, Acanthoscaphites* and related genera that make up the entire family Scaphitidae—a Cretaceous family of oddly coiled ammonites that seems to have been collectively flourishing just when the Cretaceous extinction hit.

So much for extinctions: they are very much real, ecological events that wipe out entire hordes of species of radically different degrees of genealogical affinity. When a whole family goes, it goes because all its species are eradicated. And the extinction of a species, it must be admitted, is a function of the deaths of all its member organisms—dwindling down, perhaps, like James Fenimore Cooper's last of the Mohicans, if not all cut off in the midst of their prime by an asteroid or a nuclear holocaust.

So species, everyone agrees, eventually cease to exist. There is but one cloud that confuses the issue a bit: some paleontologists speak of *two* sorts of extinction—extinction pure and simple ("without issue"—what I would call "proper" extinction) and "extinction by transformation" (also called "pseudoextinction"). George Simpson, for example, has maintained that both forms of extinction take place. It is the latter form—extinction by transformation—that muddies the waters in our search for an analogy between organisms as individuals and species as individuals. By "extinction by transformation" Simpson literally means that a single lineage (his concept of evolutionary species) continues to evolve, which is to say, to change in its anatomical and underlying genetic properties to such a degree that a former state—which we call the ancestral species—no longer exists. It has changed into something else, some descendant state, thus into some descendant species. This is nothing more than the good old gradual Darwinian-style transformational evolution, by now a familiar theme. Species, in this light, can be only anatomical abstractions, by no means the discrete sort of affairs—the *individuals*—that organisms are. After all, no organism ceases to exist because it becomes transformed into some other creature. Certainly mammals and rosebushes don't behave that way. There are many creatures that do undergo radical transformation through their life cycles: butterflies, for example. Caterpillar to winged beauty is quite a transformation. One might even wish to argue that the caterpillar stage is in some real sense a *different* individual from the adult, butterfly stage. But all are stages, however drastically different, of the same creature grown from a single fertilized egg.

Perhaps a better analogy is with the more gradual transformation from a fertilized egg to an adult mammal, like ourselves. After birth, when the transformations are less radical than *in utero*, we nonetheless change, gradually, inevitably, inexorably through time. Yet we remain a single individual. Simpson's (1961) definition of an evolutionary species reflects this understanding. Such species are, he says, "a lineage (an ancestral–descendant sequence of populations) evolving separately from others and with its own unitary role and tendencies." Thus in his definition, Simpson expects *change*. But the lineage remains one single species. It was only in practice (and indeed within the pages of the same book) that Simpson tells us we must arbitrarily chop up the gradually transforming lineages to recognize degrees of transformation—and species become stages along an evolutionary stream, not discretely top-and-bottom-bounded entities.

Thus Simpson's discussion (and all those of similar bent) poses a problem: we need discrete tops, discrete ends for species—if they are to be likened to organisms as "individuals." Simpson acknowledges the ubiquity of conventional extinction—simple termination —but was speaking of something else with his pseudoextinction. Adopting his formal definition of a species as a lineage, though, allows all manner of change to accrue without forcing us to chop that lineage up to name arbitrarily chosen levels as "species." That's the view I'll pursue here—though it really is not much of a problem after all. And in so saying, I am not being arbitrary myself in rejecting Simpson's extinctions "by transformation." The vast majority of species in the fossil record change so little internally over millions of years that when they do disappear, that's it: they've become extinct. And when they are abruptly replaced by what looks like a descendant, as *Phacops milleri* was replaced by *P. rana* in the Midwest Devonian, when the data are good it always turns out that it is *not* a case of "extinction by transformation," but rather a case of ecological replacement of one species by another—the evolutionary event connecting the two having taken place long before. Extinction means termination, just that. And that is the fate of all species. Just as it is the fate of organisms.

THE ORIGINS OF SPECIES

If "extinct by transformation" beclouds the picture of a species' end, the view that new species arise by such simple and direct transformation is equally murky. For what then can species be but landmarks along the road that a lineage treads in its persistent quest for adaptive change? I have already mentioned how Darwin saw the fixity of species, hence their reality, as the total antithesis to evolution, to the "transmutation of species." The highly respected British philosopher William Whewell had written in 1837 (only a year after Darwin's return from the five-year *Beagle* voyage) that "Species have a real existence in nature, and a transition from one to another does not exist." To establish the very idea that life has evolved as a valid scientific principle, Darwin accepted the validity of the association of the two halves of Whewell's pronouncement—and thereupon rejected the entire statement. It was only after the modern concept of species—the view that species are discrete and reproductively coherent communities of organisms—began to crystallize (in the 1920s and '30s) that it began to dawn on biologists that Darwin hadn't, in fact, literally addressed species origins in his *On the Origin of Species*. Theodosius Dobzhansky defined species as "that stage of [the] evolutionary process at which the once actually or potentially interbreeding array of forms becomes segregated in two or more separate arrays, which are physiologically incapable of interbreeding" in his *Genetics and the Origin of Species* (1937). Ernst Mayr, shortly thereafter, wrote *Systematics and the Origin of Species* (1942)—the book that perhaps more than any other work established what Mayr called the "biological species concept" firmly in the collective mind of the biological community. Mayr wrote: "It is thus quite true, as several recent authors have indicated, that Darwin's book was misnamed, because it is a book on evolutionary changes in general, and the factors that control them (selection, and so forth), but not a treatise on the origin of species." To Mayr, the reason "It is thus quite true" was simply that "Any pronounced evolutionary change of a group of organisms was to him [Darwin] the origin of a new species."

But there was, and is, no reason to accept both halves (or either, for that matter) of the good philosopher Whewell's statement. Species can be real and still give rise to descendant species—organisms, for example and by simple analogy, are doing it all the time. Mothers (not to mention fathers) are very much themselves (on the whole) after producing offspring. Darwin, for the most part, stuck with the thesis that to posit the mere existence of species was to affirm their eternal fixity and to accept the Judeo-Christian tradition of special creation by the Deity of all living "kinds." Though it took nearly 100 years fully to show that evolution need not require that species do *not* exist, nonetheless the view persists to the present day, particularly among biologists whose primary concern is the anatomy of the genetics of organisms and populations of organisms. It is their job to study the individual trees in the forest, alone and as a collective group. That the forest itself is bounded, has a history partly independent of that of other forests, is a phenomenon at the outer limits of the practical resolution of the field of genetics—though starting with Th. Dobzhansky, some geneticists have indeed been concerned with the overall gross organization of species, and their mode of origin as well.

The notion that species might be organized into collective entities and themselves be important to a consideration of the evolutionary process has always been associated with the study of geographic variation. Moritz Wagner, a German biologist and a contemporary of Darwin's, began emphasizing the importance of geographic isolation in the development of evolutionary diversification as early as 1869, a scant decade after Darwin published his *Origin*. Indeed, Darwin emphasized the importance of isolation, and utilized the tendency for closely related species to "replace" one another in adjacent regions as an argument in support of the very idea of evolution. The notion that one species stands in place of another (they were called "vicars" in good, standard Victorian usage) carries the underlying implication that species indeed *do* exist, and can bear at least a spatial relationship to one another. Mule deer *(Odocoileus hemionus)* and white-tailed deer *(O. virginianus)* generally replace each other in North America in such a fashion. J. T. Gulick and, a bit later, G. J. Romanes followed Wagner's lead in emphasizing the importance of geography to the formation of new species and emergence of evolutionary novelties—new, or modified, structures and behaviors—Romanes writing, "without isolation or the prevention of interbreeding, organic evolution is in no case possible." According to F. H. T. Rhodes (1983), Wagner began to have an impact on Darwin that was reflected in some of Darwin's emendations to the *Origin* in later editions.

But geographic variation involves adaptive features of the organisms—anatomical, physiological and behavioral items—and isolation was seen to encourage the development of such diversity. Thus geography was seen to play the same role as time in the evolutionary process: given enough *space* (just like time), anatomical change is likely, perhaps even inevitable. Species remain no truly special thing—just early phases in a process of anatomical differentiation, anatomical transformation in space as well as through time.

Mayr's "biological species concept," in its shortened form, is still memorized by students. Species are "groups of actually or potentially interbreeding natural populations, which are reproductively isolated from other such groups" (Mayr, 1942, p. 120). Mayr—and Dobzhansky—saw species as reproductively coherent communities. And although some species could hybridize with others (lions and tigers, for example), generally such reproductive communities are discrete: generally speaking, members of one sexually reproducing species mate only with other members of the same species. Mistakes are relatively rare.

And yet no two species are identical. No two organisms within a species—save identical twins—really are identical. But there would be no way for there *not* to be interbreeding between "species" unless there were some collective differences between the members of the two groups—some way for organisms to recognize members of their own species, and not to mate with members of other groups. According to biologist H. E. H. Paterson, such mate-recognition cues might be visual and thus obvious to us as "species markers." Or the cues may be subtle chemical differences, preventing ambient sperm in seawater from fertilizing any eggs but those of a conspecific female, though the sperm may contact the eggs of any number of other species of invertebrates living together in a seafloor community. There is a network of reproduction interlinking the members of a species, and it is indeed true that the very need to recognize mates of one's own species tends to keep species apart and to ensure that there will be some set of anatomical, behavioral or biochemical properties that mark species as distinct, as different one from another.

So there *is* a link between "transformational" properties of organisms and differences between species. Darwin's tradition has focused solely on this link, to the point of seeing "evolution" synonymous with "transformation of features," and the latter synonymous with "speciation." But adding the further ingredient of separate clusters of reproductively coherent communities—species—enables us to look at the problem a bit differently. It turns out that there is an enormous range of visible anatomical differences between closely related, reproductively isolated species. Some closely related species are so similar that biologists for years never knew they had two different species in their collections. Fruit flies, the darlings of genetics research (because their salivary glands house cells with giant chromosomes, and because fruit flies are a tractable laboratory creature that breeds rapidly), have produced some of the more famous examples of such "sibling species"—such as the species *Drosophila persimilis* and *D. pseudoobscura,* discovered to be different only when Dobzhansky found greatly reduced ability to interbreed between two groups that had been called "races A and B" of *D. pseudoobscura.*

More typical is the situation depicted by Mayr, entailing subtle, yet distinct differences between closely related species. Mayr wrote about five species of common eastern North American thrushes, all of the genus *Hylocichla.* (Or at least, they were all attributed to that genus as Mayr wrote. Increase in our understanding of the relationships among species often leads to species' being reassigned to other groups, which prompts a name change as well, to the frequent consternation of those of us unaccustomed to flightiness in the names by which we call familiar things. Nowadays only the wood thrush remains in the genus *Hylocichla.* All the rest are now in the genus *Catharus*). Mayr wrote:

> If we examine the variation within the genus in more detail, we find that it clusters around five means, to which we apply the familiar names wood thrush *(Hylocichla mustelina),* veery *(H. fuscescens),* hermit thrush *(H. guttata),* gray-cheeked thrush *(H. minima),* and olive-backed thrush *(H. ustulata).* All five species are similar, but completely separated from one another by biological discontinuities. Every one of the five species is characterized not only morphologically, but also by numerous behavior and ecological traits. Two or three of them may nest in the same wood lot without any signs of intergradation; in fact, not a single hybrid seems to be known between these five common species. I could list genus after genus of familiar North Ameri-

can or European birds and demonstrate exactly the same [Mayr, 1942, p. 148].

And indeed Mayr's description of a number of fairly similar, yet always discrete, closely related species living in the same neighborhood is a common occurrence in nature. And, close as some of those species might at first glance seem, there always is that list of anatomical and behavioral features which allows us to tell them apart.

Some closely related species, of course, seem *very* different from one another. Lions *(Panthera leo)* and tigers *(P. tigris),* for example, are still sufficiently compatible genetically to form hybrids—as they have done, for example, in zoos, with "tiglons" the unfortunate result. Tigers are a bit larger than lions, and of course distinctively striped—quite a bit different from the monochrome tan lion, whose males are typically distinctively maned. The social organizations of the two species are even more markedly different: lions are organized in prides (with some loner males), while tigers, usually living in denser vegetation than lions, go it pretty much alone (males—and females once the cubs are old enough). Yet without their skins, by bones alone, competent anatomists tell lions from tigers, if at all, only with great difficulty.

So there is a spectrum. Reproductively isolated groups, true species, may differ a great deal, or hardly at all, one from another. Both Dobzhansky and Mayr stressed this dual nature of the evolutionary process: diversification, through the Darwinian-style transformation of the anatomical and behavioral attributes of organisms, and discontinuity, or those gaps which interrupt that anatomical and behavioral diversity, gaps which coincide precisely with reproductive discontinuities. The gaps set off those reproductive-*cum*-anatomical packages which we call "species." The explanation of the diversity—natural selection modifying the anatomies and behaviors of organisms from generation to generation as an adaptive response to new or changing environmental conditions—seemed to Mayr and Dobzhansky to be secure. And so it seems today. Throughout a species' range, organisms vary from place to place as they become adapted to somewhat different ecological conditions—the common phenomenon of adaptive geographic variation. But what is the origin, then, of the discontinuities between species?

The clue, once again, lies in geography. Mayr, especially, pointed out that any two species living at the same time in the same place (they are "sympatric") would *have* to be firmly distinct one from the other, whereas we might expect there to be more of a problem telling species apart as we go from region to region, trying to decide whether a variant form belongs to a closely related, but distinct, species or is instead a member of a single widely varying species. Allopatrically (that is, with the organisms living apart) anatomical gaps should be less pronounced than those which occur when the organisms are sympatric. Mayr writes (1942, p. 149): "The gaps between sympatric species are absolute, otherwise they would not be good species; the gaps between allopatric species are often gradual and relative, as they should be, on the basis of the principle of geographic speciation."

Thus, to Mayr, sympatric species are always easy to spot, and there is no doubt that they are "real." Allopatrically, species pose a problem as they intergrade from place to place. In an important passage, Mayr writes of the practical difficulties in dealing with allopatric species—nonetheless affirming his conviction that species are *real* entities, rather like the individuals as which Hull and Ghiselin would have us see them:

> In addition to these clear-cut and "bridgeless" gaps between sympatric species, there are, however, the gaps between allopatric forms, and the unbridgeability of these gaps is often very doubtful. Nearly every well-isolated population which has developed some characters of its own may be considered a separate species on the basis of certain criteria. The decision as to whether to call such forms species or subspecies is often entirely arbitrary and subjective. This is only natural, since we cannot accurately measure to what extent reproductive isolation has already evolved. In fact, such cases are logical postulates, if the divergence of isolated populations is one of the important means of species formation. A species evolves if an interbreeding array of forms breaks up into two or more reproductively isolated arrays, to use Dobzhansky's terminology. If we look at a large number of such arrays (that is species), it is only natural that we should find a few that are just going through this process of breaking up. *This does not invalidate the reality of these arrays; just as the* Paramecium *"individual" is a perfectly real and objective concept, we find in most cultures some individuals that are either conjugating or dividing* [Mayr, 1942, p. 152; emphasis added].

Thus Mayr unequivocally accepts the first half of Whewell's statement—species are real—yet obviously rejects the second. To Mayr, life has evolved, and in particular, new species arise from old. Species "transmutate," all right, but by budding off descendants, and *not* by giving up the ghost and becoming transformed into wholly new, descendant species.

Or so, at least, it seems. Such *is* the substance of Mayr's views; yet, as we have already seen in a passage cited earlier, Mayr went along with virtually all the rest of evolutionary biology with his picture of the nature of the existence of species through time: once a species evolves, it will inevitably continue to evolve itself out of existence.

And it is this latter view, with its Darwinian roots, to which Gould and I take exception. Species seem to us very much as "real" in time as Mayr himself saw them in space. But before we apply these principles to the fossil record, we still must ask how those reproductive discontinuities—hence anatomical discontinuities—become established in nature.

How, in fact, *are* reproductive communities disrupted in nature, sundered to yield two (or more) species from a single ancestral species? Dobzhansky, Mayr and most biologists after them have had difficulty imagining any ultimate source of reproductive discontinuity other than the by-now-familiar geographic isolation: simple physical disruption of a once continuously distributed species. In point of fact, the matter is not so simple. No species, even our own, is totally continuously distributed. Most species are broken up into small populations; many of these little "colonies" seldom if ever contact other groups. The granite kopjes that stick out from the grasslands, dotting the plains of eastern and southern Africa, sport their own little ecological communities. Members of some species, such as lions and leopards, will move freely from kopje to kopje, while the little klipspringers (antelope adapted, like our mountain goats and sheep, to life on rocky crags) are pretty much stuck where they are. Few things in nature are inevitable, and physical isolation need not always trigger reproductive isolation. Yet reproductive isolation, it seems, seldom if ever becomes established *without* geographic, physical isolation disrupting a species.

The idea is simple enough, and seems to fit patterns of distribution of organisms in nature pretty well. Given physical isolation, particularly of relatively small populations near the edge of a species' range, continued adaptive divergence of a population (through natural selection) may produce enough modification that, should the isolate ever come into contact again with the "parental" species, the organisms either will not recognize each other as potential mates, or will try to mate and fail—all because they have effectively drifted apart, gone their separate evolutionary ways long enough that the genetic change accumulated within each group will be too great to permit successful fertilization or the development of reproductively fertile offspring.

Mere isolation, plus a sort of "single-mindedness" leading to adaptive specialization of the isolated population, may lead *fortuitously* to the accumulation of enough genetic differences simply to make the two groups—parental and fledgling daughter species—incompatible. It is as simple as that. Geographic isolation *may* lead to the accidental formation of new species by the mere accumulation of genetic differences in the two segments of a single ancestral species.

Yet arguments, as they will, persist. The details of the genetic differences that prevent interbreeding are numerous and sundry. Some instances of speciation are the result *not* of small populations near the periphery of a species' range being cut off, but rather by the fragmentation smack down the middle of a single species. When, for example, the Isthmus of Panama rose above sea level most recently (some 4 million years ago), cutting off Caribbean sea dwellers from those in the Pacific, many species were divided in two. By now a number of species seem to have fragmented into distinct groups on each side—sufficiently different to warrant being called distinct species by some biologists (though we should always bear in mind Mayr's discussion of the logical and practical difficulties of telling whether two allopatric groups are "true" species or not). The point about the ability to interbreed is relevant here: one could transport members of two similar clam species from the Caribbean and the Pacific to a lab and see if they are interfertile. But the point of the "reproductive criterion" for delimiting species in nature is whether or not the two groups actually do interbreed in nature—not what they could do if introduced to each other in the lab or at a zoo. Lions and tigers *are* interfertile in zoos, but do not hybridize in nature (they *do* still co-occur to some limited extent, in India). That's what Mayr was getting at: when two populations are separated in space (or, for that matter, in time) we should ask not whether they could interbreed if they were brought together, but whether there is a full skein of reproductively interfertile members of the species that in fact link up the groups over space and through time. Species, as Mayr said, will always be difficult to identify in practice—except when they are living in close proximity and boundaries between them are usually clear to all.

Other biologists have taken issue with the generalization (again, largely due to Mayr) that the origin of reproductive isolation must inevitably be found in enforced geographic isolation. From time to time biologists have advanced new theories explaining how reproductive isolation can become established *sympatrically*—within an ancestral species' range. Most arguments for a behavioral or ecological shift on the part of some organisms within a species—a shift that would keep them from mating with the rest of the members of the species—have failed to prove convincing. More plausible are the notions of relatively large-scale chromosomal reorganization, such as those documented by Australian biologist Michael White and colleagues, working on flightless grasshoppers. White postulates that chromosomal races can develop "parapatrically"—that is, in closely adjacent areas where the boundaries between groups are sharply defined. Total disruption of intergroup reproduction by such chromosomal rearrangements are *ipso facto* quick, leading to "parapatric" speciation. The trend these days is to relax a bit from the credo that absolutely *all* speciation need be allopatric—yet it is equally a valid generalization that most analyzed cases seem to fit the picture of allopatric speciation best.

Thus speciation is very definitely something more than the simple transformation of the anatomical features of organisms. Species can vary widely over their territories, and can change much through time, without breaking that all-important reproductive network, that complex chain of interfertility which imparts cohesion to a species. Indeed, I have elsewhere referred to this reproductive cohesion as the "glue" that binds a species together—the final element needed for a complete analogy between organisms-as-individuals and species-as-individuals. Our bodies are literally bound together by cellular cohesion. Our skin defines our outer limits, but cannot alone hold us together. Species are bounded in space and time and are likewise held together by a very real source of internal cohesion: that reproductive plexus. Break it long enough, and you are likely to end with two species instead of one. This picture of nature's evolutionary dynamic is quite a bit different from Darwin's and that of his tradition: it is an overlay of fortuitous, ecologically based reproductive isolation on the more traditional story of adaptive transformation. And, seemingly, it makes a far better explanation of the data of the fossil record than the equally more traditional picture that pure Darwinian "transformationalism" affords.

BACK TO THE FOSSILS: THE ORIGIN OF PUNCTUATED EQUILIBRIA

Phacops rana and its kin posed a devilish problem. Once some change finally did emerge within the lineage over its 8-million-year history (about the same order of anatomical change as Mayr described for those five species of eastern woodland thrush), the patterns still did not fit the picture of gradual transformation of an entire lineage that zoologists and paleontologists alike had always expected to find in the fossil record. Once again geography seemed to be the key: the little cow-pasture exposure in the extreme northeastern margin of that ancient Devonian inland sea, with its hard-to-find assemblage of trilobites that seem to bridge the gap between ancestor and descendant, in particular seemed to cry out for some sort of alternative explanation. The apparent, and apparently sudden, transitions in the Midwest were phony: the real transition had occurred millions of years earlier, seemingly only in one corner of the ancestral species' range.

It was only natural to take that pattern and match it up with an alternative set of explanatory evolutionary theory. There was no need to invent some sort of fanciful alternative. Allopatric speciation seemed a process ideally suited to explain all that I had found had happened in the 8-million-year history of the *Phacops rana* group—all, that is, except for the rather astonishing degree of nonchange, stolid stasis, that if anything seemed the predominant theme of the group's entire history. But it wasn't just a *faut-de-mieux* choice and a desire to be conventional: allopatric speciation continues to look good as a reasonable explanation for the origin of the patterns of change that *did* occur within that trilobite lineage. Those species—

Phacops milleri, P. rana and, finally, *P. norwoodensis*—according to this interpretation are to be taken literally as true biological species. How one can defend such an interpretation given the obvious dearth of information about the reproductive proclivities of organisms dead for 375 million years is a matter we'll examine a bit more closely in following chapters.

I wrote up the *Phacops rana* story in a short article published in the technical journal *Evolution* in 1971. Under the rather ponderous title "The Allopatric Model and Phylogeny in Paleozoic Invertebrates" I was simply telling the tale and drawing the homily that perhaps allopatric speciation was more apt than wholesale phyletic lineage transformation as a biological process underlying patterns of stasis and change in the fossil record. In particular (and I was able to cite a few paleontologists in my support) I claimed that stasis (I called it simply "stability") was a general phenomenon in the fossil record, that new species typically appear rather abruptly in the fossil record, that they commonly overlap in time with their apparent ancestor and that if we looked hard enough, we would more than likely be finding new species arising in apparent geographic isolates (as I seemed to have found with *Phacops rana*)—in other words, life was speciating back then just as we believe it is doing now. Uniformitarianism in action.

Meanwhile S. J. Gould had (several years earlier) completed his dissertation on the evolutionary history of the Pleistocene (Ice Age) snail *Poecilozonites*. Gould—who had a complete evolutionary "microcosm" on hand, as those snails were absolutely restricted to Bermuda—documented the same sort of abrupt appearances and stasis that I was finding in the trilobites. The origin of "punctuated equilibria" (Gould's term) was a natural partnership. Paleontologist T. J. M. Schopf asked Gould to contribute a paper on speciation and the fossil record for a symposium-*cum*-book, *Models in Paleobiology*—a project expressly geared to inject some interest in biological theory within the ranks of paleontology. Gould accepted, though speciation was not his first-choice topic (he would have preferred to write on theoretical aspects of growth and morphology—among the many areas of his active interest. But those topics had already been spoken for in the symposium plans). Gould simply asked me to write the paper with him, as I had by that time produced the *Evolution* manuscript and he felt the same basic way about the problem of explaining evolutionary patterns in the fossil record. The topic seemed promising, as what we offered was an alternative way of interpreting paleontological data.

I have reprinted the original punctuated equilibria paper as an appendix to this book. It was first published in Schopf's (1972) *Models in Paleobiology*, pp. 82–115. Gould and I, after much discussion, each wrote roughly half. Some of the parts that would seem obviously the work of one of us were actually first penned by the other—I remember, for example, writing the section on Gould's snails. Other parts are harder to reconstruct. Gould edited the entire manuscript for better consistency. We sent it in, and Schopf reacted strongly against it—thus signaling the tenor of the reaction it has engendered, though for shifting reasons, down to the present day. More of all that anon.

At base, punctuated equilibria boils down to a few simple propositions. Today it is its implications that provoke controversy; in 1972, it was the amalgam of these simple propositions that fanned some flames. At its simplest, punctuated equilibria entails the recognition of stasis and the realization that patterns of change in the fossil record are best explained by allopatric speciation. Stasis was known to paleontologists before Darwin (all the paleontological critics of the *Origin* mentioned stasis in their reviews!)—though stasis had become something of a professional embarrassment to be politely ignored, so alien did it seem to what evolution ought to look like in the fossil record. In a very real sense Gould and I merely resurrected stasis as an empirical fact of life and argued that this prodigious lack of change must be accounted for every bit as much as change itself in any theory of the evolutionary process. Stasis remains an interesting phenomenon for evolutionary biologists to chew over, and we'll take a harder look at it in the next chapter.

Allopatric speciation, equally, was not of our making. Mayr had years earlier refined the concept, but its roots go back through Dobzhansky and indeed far back into pre-Darwinian biology. In 1954, toward the end of a paper in which he discussed the importance of peripheral isolates in speciation, Mayr had even suggested that such peripherally isolated populations could go a long way toward explaining gaps in the fossil record—the apparently sudden appearance of new species which might simply be migrating in from that periphery.

But Darwin had effectively said that adaptive change was the origin of species. No one had ever claimed just the opposite—that most (if not all) anatomical change in evolution, adaptive though it may be, happens not throughout the bulk of a species' history, but rather at those rare events when a new reproductively isolated species buds off from the parental species. Speciation acts as the trigger —not the result—of adaptive change. Most speciation events culminate in rather modest degrees of change; minimally, only some modification of the reproductive system is required for speciation to occur. (It is a persistent misconception that we claimed that every speciation event necessarily entails a major degree of anatomical alteration: most speciation events, as we have seen, involve truly modest degrees of change.) Or so we felt the fossil record was telling us: for that is the simple conclusion if we use allopatric speciation as a process to explain the pattern of episodic change we saw as so very common in the fossil record. Nor were those "abrupt" events truly ecologically instantaneous: 5,000 to 50,000 years became the comfortable yardstick, compatible with the admittedly rather coarse level of resolution in our own paleontological data, and compatible as well with the time scales required in most theories of the speciation process.*

Writing primarily for paleontologists, we were simply advocating substitution of one set of evolutionary principles for another as a way of addressing the all-important problem of fitting fossils into some sort of coherent evolutionary framework. We were not prepared for the reaction. Whether favorable ("I knew it all along" or the more positive "Finally I can make sense of my data") or critical ("You guys are full of ———" heard more than once at national meetings in the mid- to late 1970s), the reactions provoked by the paper were strong among nearly all paleontologists—simply because most paleontologists had some personal experience with the fossil record that they could match up with our version of reality. Some paleontologists were angered that we implied that the profession as a whole was ignorant of the simple model of allopatric speciation (our reply: of course we all know about it, and even teach it to our students—but we seldom if ever apply it in our research!). Other criticisms were more serious—including the charge that one cannot see biological species in the fossil record, and the denial that

* The assertion that punctuated equilibria represents a resurrection of Goldschmidt's "macromutations" and "hopeful monsters" remains the most serious and irksome misconstrual of our ideas.

stasis is a general phenomenon—or even occurs at all! More of these and related matters in the next chapters.

But, as I hinted earlier, there was more to the basic model of punctuated equilibria than the amalgam of empirical stasis and the explanation of episodic change as the simple and expectable consequence of allopatric speciation. For punctuated equilibria puts the icing on the cake in the argument that species are real historical entities, comparable in a formal manner to individual organisms. Toward the end of our 1972 paper, Gould and I posed a paradox, a strong challenge to the unfettered efficacy of natural selection both as a sweeping cause and as a *modus operandi* of all evolutionary change. We began to see species more as independent actors themselves in the evolutionary arena. That's the part which eventually provoked reaction (most of it hostile) in other spheres of the evolutionary-minded biological community. And that's where most of the fun lies in the 1980s.

5 PARADOX FOUND: ADAPTATION AND MACROEVOLUTION

THERE ARE some thirty-eight active volcanoes on Java. Sitting on the edge of a subduction zone, where oceanic crust is being sucked down and heated under the major islands of the Indonesian archipelago, the volcanoes are vents for some of that deep-seated mélange of molten rock. Though much of Java is now dedicated to rice cultivation, the fuming volcanoes and lush greenery still present a distinctly primitive feel. And one wonders why *Homo sapiens*, in all its rich cultural and ethnic diversity, is there at all. What ever happened to Java man?

"Java man" was the discovery of a Dutch physician, Eugène Dubois. Taking his cue from Darwin's *Descent of Man,* Dubois wanted to go to Africa to search out the fossilized remnants of man's early relatives. Darwin had predicted that Africa would prove to have been our cradle simply because our closest living relatives—chimps and gorillas—are themselves uniquely African. But Dubois couldn't afford Africa, and had to settle instead for enlisting in the Dutch East India Army. In 1887 he sailed for Sumatra, and by 1891 he had found his first human fossils—a tooth and a skullcap—in central Java. These were the first of what is by now an impressive array of fossils that span an interval of nearly 1.5 million years—and that occur from Indonesia through Asia and Europe and on down to the southernmost reaches of Africa. Dubois dubbed his Java man *Pithecanthropus erectus* ("erect apeman"), while specimens recovered from a cave near Peking in the 1920s and '30s were "Peking man" *(Sinanthropus pekinensis).* By the 1960s paleoanthropologists were calling all such fossils *Homo erectus,* which signals their recognition that all these diverse fossils belong to Dubois's species (*erectus,* the first to be named). And referring *erectus* to the genus *Homo, our* genus, was admitting that this primitive hominid really was our very close relative.

Now, *Homo erectus* is fascinating in several ways. It is the sole known hominid species throughout the Middle Pleistocene. In its earliest days it was a contemporary of some of the robust and very primitive species of *Australopithecus,* a collateral relative, in Africa. But the australopithecines became extinct slightly more than a million years ago, and *Homo erectus* alone was left among the hominids. In a gross *Gestalt* sort of way, *H. erectus* is a perfect "intermediate," what used to be called a "missing link": when skulls of the primitive australopithecines are lined up at one end of a table, modern human skulls at the other end, and *erectus* placed in the middle, everyone (save a creationist) will readily agree that *H. erectus* falls anatomically somewhere between the two extremes. With its heavy eyebrow ridges and flattened forehead it recalls the australopithecines—and further back beyond them, the great apes themselves. But the average brain size of the *erectus* specimens hovers around 1000 cubic centimeters, with extremes of 800 to more than 1200 c.c. Australopithecine brains average around 400–500 c.c., while the forerunner of *erectus, Homo habilis,* comes in at 775 c.c. In contrast, the average brain size of modern *Homo sapiens* is 1300–1400 c.c. *Homo erectus* falls midway in time and anatomy between the older, more primitive hominids on the one hand and ourselves on the other. Could there be a better example of evolutionary change from the fossil record?

Not really. We tend to think of evolution simply as change through time—and there it is in spades in our own lineage. *If* the general notion that life has evolved is correct, then it simply must be true that all species, dead and living, are interconnected in one vast kinship network. All living species, including ourselves, must be most closely related to some one particular species of all those now sharing the globe. Darwin's candidate for our closest relative was one of the species of African ape (though J. H. Schwartz has recently provided good evidence that orangutans, which are native to Borneo and Sumatra, are really our closest living relatives). Whichever ape *is* our nearest kin, when we compare ourselves with any ape it is obvious that both relatively and in an absolute sense, we have the bigger brains. Ergo: in the course of human evolution, we expect to see a progressive increase in brain size from apish figures to modern volumes as the fossils get younger and younger. And, as I sketched above, this prediction is exactly borne out by the fossil record.

Homo erectus is the midpoint of a *trend* in human evolution—a trend toward increase in brain size. There were other such trends involving overall body size and other, more subtle changes. Brain size, though, is the easiest to see and easily the most arresting of all features of human physical evolution: for in some gross way, at least, brain size has something to do with intelligence, and we can readily imagine, just looking at ourselves today, that our capacity for speech, complex social organization and the ongoing invention and use of a sophisticated technology *is* our species' basic adaptation—a capacity rooted somewhere, somehow in our brains. The archeological record bears this point out nicely, too: just as the bones get progressively more modern through time, so too do the ancient technologies, while size and complexity of social systems increase as well.

Human evolution, then, seems an absolutely perfect example of progressive, adaptive evolution, the sort of regular, protracted change Darwin thought we ought to encounter routinely in the fossil record. And that's just the way human evolution *is* usually interpreted: slow, steady, gradual, progressive change, from our most apish ancestors on up through intermediate forebears, such as *Homo erectus*—and on and on, till we ourselves emerge some 50,000 years ago. And human evolution is so readily interpretable in Darwinian terms of natural selection: the "selective value" of progressive increase in brain size seems so patently obvious, the adaptive worth of ever-more-complex intelligence so unarguable, that it hardly has occurred to anyone to question the application of these simple Darwinian axioms to the fossil record of human evolution. And as I have briefly presented the facts so far, there seems to be little reason to doubt the story.

But there is a fly in the ointment—and once again we confront *Homo erectus*. Yes, *erectus* is a perfect anatomical and temporal intermediate in the long saga of human evolution. But, like *Phacops rana*, which seemed to stay so perfectly still (in evolutionary terms) for such prodigious periods of geological time, *H. erectus* is a wonderful example of a successful, stable species. Philip Rightmire (1981) has analyzed all the available data on *erectus* and concluded that though the brains of the earliest known specimens of *H. erectus* were, on average, a bit smaller than the youngest known specimens, the differences are statistically insignificant. And some of the earliest specimens—such as Richard Leakey's ER-3733 from Lake Turkana in eastern Africa, with its brain size of 850 c.c.—are astonishingly similar to the classic *erectus* specimens from China, specimens that are younger by more than a million years. And once again the archeological evidence of the culture of *Homo erectus* (scanty though it is, as it consists almost solely of stone implements) likewise offers a picture of great stability. For more than a million years, the so-called "Acheulian" industry, consisting of bifaced handaxes struck from a stone core, persisted with little change in Africa, western Asia and western Europe (where the bones of *H. erectus* have scarcely been found). There is more variation, it seems, geographically in the precise stylistic developments of these handaxes than there is any pattern of concerted change through time. Indeed, the skeletal remains of *erectus*, so recognizable from Africa to Indonesia, are actually more uniform than the cultural implements: bifaced handaxes struck from a stone core are not the rule in Asia, where the more primitive chopper (a crudely shaped implement fashioned from the rock core itself) persists as the tool tradition of *H. erectus*, a cultural holdover from remoter times.

Geographic variation, but no concerted change for nearly a million and a half years, is the picture we get of both the anatomical and cultural histories of *Homo erectus*. Stasis is the rule. And *H. erectus* really seems like a successful, far-flung species, regionally differentiated to some extent and, with its particular level of cultural development (whatever the details may have been), admirably suited to survive in a wide spectrum of climates and habitats in mid-Pleistocene times in the Old World. The picture we begin to get whenever we look at these "intermediate" species—these *Homo erectus, Phacops rana* and the like—is hardly one of a passive, transitory and transitional phase somewhere in the midst of an evolutionary stream. What we see is a widespread, successful species coping very well with life's exigencies, able to persist for millennia. And looking at them this way, we begin to ask an entirely new set of questions: How come this stability? And what is the role of these stable species in the grander scheme of things? Given the inevitability of environmental change, where the old Darwinian view logically presumed evolutionary change to be inevitable, we now see that it is not, and we ask why. The success—without change—of *H.*

erectus was so complete for so long that we come to wonder why *Homo sapiens*, and not *Homo erectus*, occupies Java today.

And here is the paradox that Gould and I saw in 1972. It *is* true that long-term evolutionary change frequently amounts to protracted, *directional* change—brain size in human evolution giving us a perfectly typical example. The usual explanation: what we see in laboratory experiments in genetics, where artificial selection produces generation-by-generation change, is assumed to be a microcosm of what goes on in nature. As we have seen, the thought is simply that given enough time, slow, steady, small-scale change can add up to quite a lot. Given enough time, all manner of change can and will happen. And this leads to the view that species are but a passing stage in the evolutionary stream. In other words—and here is the key credo of modern evolutionary theory—natural selection is by far the most important process acting within species. Selection works on the genetic variation expressed in the features of organisms within populations and species. And all the truly large-scale phenomena of evolution—what we call "macroevolution"—can be understood as a simple summation, a simple extrapolation, of the within-species process of natural selection. Evolutionary trends are a profound case in point: *modern evolutionary theory sees evolutionary trends as the large-scale accumulation of directional natural selection.* Within-species processes and longer-term processes are all the same: all we need consider is the generation-by-generation process of natural selection.

But *Phacops rana* kept a stable 17 columns of lenses in its eyes for about 6 million years, though it was the intermediate in a sequence of an 18–17–15 trend in column reduction. *Homo erectus* was an intermediate in a series of fossil hominids with progressively larger brain sizes—a trend thought to be produced because naturally natural selection will favor increased brain size for its greater realized intelligence. Yet for a million and half years selection did no such thing—and never before (to judge from the ubiquity of *erectus* in the Old World and its invasion, the first for hominids, into northern temperate climes) had there been such a successful species in the entire history of the human family. It is the old argument taken up a step. So far, the *Phacops rana* story of stability and speciation tells us that the expectation that evolutionary change is slow, steady and gradual is at the very least not always borne out by the facts. But if we deny, simply on the evidence of our senses, that gradual progressive change is the rule, the business-as-usual nature of the evolutionary process, we have also called into question the favored explanation for all large-scale evolutionary phenomena—the most egregious being directional trends. For the paradox is this: *species typically do not exhibit the sort of directional change through their long histories, the sort of pattern that has been considered the true stuff of all evolutionary change. Yet species very commonly are members of lineages—trilobites, hominids, horses, dinosaurs—in which such directional change between species is a commonplace.* Gould and I claimed that stasis—nonchange—is the dominant evolutionary theme in the fossil record. It is characteristic of most species that have ever lived. Adaptive change is relatively rare, and usually associated with speciation, thus typically rather rapid. Once a species appears, if it is successful at all, the fossil record shows that it will tend to hang on unchanged for vast stretches of time. And this, we saw, destroyed the backbone of the major argument of the modern "synthetic theory" of evolution—the argument maintaining that absolutely all the features of the history of life could be seen as constantly ongoing adjustments to ever-changing environmental conditions. To understand this better we need to take a closer look at this phenomenon we call "adaptation."

THE ADAPTIVE LANDSCAPE

One of the time-honored arugments for the existence of an omnipotent Creator is based on the close fit between organisms and their environments. Birds have wings to fly, but eagles soar and hummingbirds hover, and their body sizes, wing shapes and endowment of feathers all reflect the specialized ways those birds actually do fly. Nature programs on television are almost solely litanies of behavioral and anatomical masterpieces of engineering, all designed to show us how admirably suited cheetahs are for catching their prey, or how adept wildebeest cows are in defending their young. The message in all these treatments is the same; only the morals differ. In the 1820s, the "argument of design" triumphantly bespoke the existence of a Creator—for just as a watch, with all its interdependent intricacies, directly implies a watchmaker, so too does the existence of natural design imply a Master Designer. The world is

If evolution within species typically involves gradual change, it would look like this:

. . . and we would have no problem seeing long-term, large scale changes, such as increase in brain size in human evolution, as looking like this:

But species are typically stable, and look more like this throughout their life spans:

. . . so, there is a paradox: despite within-species stability, we still see protracted changes within lineages. How can we reconcile large-scale change with stability within species, when we relied before on within-species change to explain large-scale trends? The answer comes from the pattern, which looks like this:

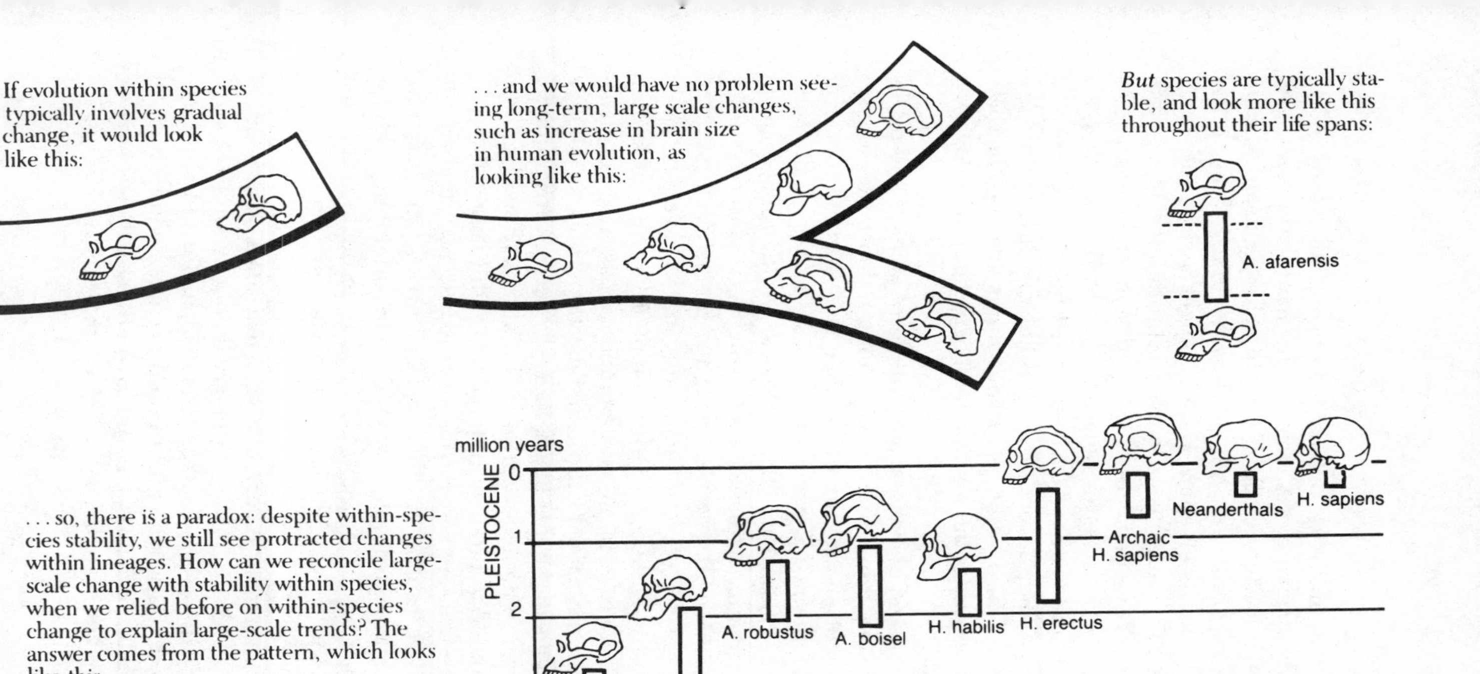

Based on drawings by N. Amorosi in Eldredge & Tattersall, 1982

amazingly ordered, and each organism so beautifully designed to carry out its role that only a Creator could have fashioned it so. Nowadays, though modern creationists still echo the analogy, biology still speaks of design but follows Darwin in ascribing to it a material cause: natural selection. Darwin taught us to think of nature's design, the fit of organisms to their surroundings, in a purely naturalistic sense. Yes, organisms are shaped to meet the exigencies of existence, and sometimes in remarkable ways. But we can see wings, say, as modified legs—derived from the walking limbs of reptilian forebears and, beyond, from the fins of fish. And we can see the diversity of bird wings as various forms adapted for some more specific use, such as the soaring flight of eagles or the hovering of hummingbirds.

The structural and behavioral adaptations of animals and plants are all *used* for something, usually something rather basic, like securing nourishment or warding off predators. The grand theme of organic nature is the maintenance of the body proper. In the complex web of energy flow that is the ecological world, each organism seeks its supply, and in turn contributes to that flow. Most of nature's design is concerned with precisely this: the niches each group of organisms occupies, the ways and means organisms have developed simply to earn a living. Success in the competitive realm of limited resources was Darwin's simple and eminently reasonable explanation of how organisms came by those modifications, how they honed those modifications in the first place. Indeed, "stasis," or stability of species, is in a sense no real surprise to evolutionary theorists, who are long accustomed to thinking in terms of "stabilizing selection," in which it is the "normal" organisms, not the extremes, that are the more successful at living—and that consequently, in the long run, end up leaving more offspring. Though Darwin had trouble with his critics, ever since the 1930s there has been little doubt within the ranks of evolutionary biology that all that design out there in nature is the product of natural selection.

Modern evolutionary theory in a very real sense is a theory of adaptation. Indeed, *Darwin's* evolutionary theory, as we have seen, was essentially a theory of adaptation, an explanation of how organisms have come to be so admirably suited to such a wide variety of methods and modes of making a living. Evolutionary theory today is essentially an updated version of the "modern synthesis." Forged in the 1930s and '40s, this synthesis marked a return to Darwin's straightforward thinking. Though he had succeeded, nearly right from the start, in persuading thinkers of the Western world that evolution had indeed occurred, Darwin ran into persistent difficulties in convincing his peers that natural selection is *the* mechanism of evolution. And when the early papers of Gregor Mendel were "discovered" (independently, by three different biologists) at the turn of the century, genetics was launched—and immediately collided with by-then-standard canons of Darwinian orthodoxy. Spontaneous change—"mutations"—appeared often to have quite large effects, and some of the influential early geneticists (particularly the Dutchman Hugo DeVries) were in effect early forerunners of Richard Goldschmidt: they claimed that evolutionary change was far more a reflection of rare, sudden genetic changes than the accumulation, by selection, of slight variants within a population. Selection was assigned merely a mop-up role, eliminating the unworkable among the mutants. To many of these early geneticists, it was difficult to see what "creative" role selection could play; in Darwin's vision, of course, it was selection that forged the anatomical and behavioral features of organisms that we call "adaptations."

Three mathematically inclined geneticists, Sewall Wright in the United States and Ronald Fisher and J. B. S. Haldane in England, were instrumental in removing most of the conceptual difficulties that the early results of genetics research seemed to pose for Darwin's theory. They pointed out that the effects of mutation are far more likely to be slight than marked; they showed mathematically that selection, even at modest intensities, could accumulate substantial change in the representation of genes within a population over just a few generations. The stage was set for a real rapprochement between Darwin's original views and the maturing science of genetics, a function fulfilled in 1937 when Theodosius Dobzhansky published his *Genetics and the Origin of Species.* Ernst Mayr soon followed with his *Systematics and the Origin of Species* (1942), and paleontologist George Gaylord Simpson soon thereafter (1944) with his *Tempo and Mode in Evolution.** The net effect: in a brief span of less than ten years, biology was seemingly brought together, once

* I have recently analyzed these books in detail in *Unfinished Synthesis* (1985). In addition, Ernst Mayr and historian W. B. Provine (1980) have edited a very useful collection of essays on the history of the emergence of the "modern synthesis"—or "synthetic theory of evolution."

again united under the all-encompassing notion of evolution, and Darwin had been abundantly vindicated in his basic scheme.

Today's dominant evolutionary theory is nothing more nor less than a refinement of the "synthesis." And perhaps nothing better captures the very essence of the synthesis than the imagery of the "adaptive landscape." In 1932, Sewall Wright published a brief paper in which he pointed out a few simple genetic truths: of all the different genes, and of all the different forms each gene could take, it stood to reason that some of the combinations of those genes would produce a more vital, vigorous organism than some other combinations would create. Wright used the word "harmonious" for those better combinations. And he drew a crude map, a topographic diagram of hills and valleys, and suggested that we could symbolize those relatively more harmonious combinations as occupying the "peaks," and the less harmonious combinations the "slopes" and "valleys," of this "adaptive landscape." Thus did Wright depict the central problem in evolution: How, he asked, does a species maximize the number of organisms on the peaks and minimize those in the valleys? And how does a species manage to move its members from some peaks relatively low in the field to those peaks which symbolize even more harmonious combinations of genes?

Wright, of course, answered his own rhetorical questions, posing a model that emphasized trial and error in the exploitation of new peaks—trial and error even over the power of natural selection. When Dobzhansky, Mayr and Simpson became the major voices in evolutionary theory in the following decade, they relied heavily on Wright's work, particularly on his adaptive-landscape imagery. And in adopting Wright, they made only two changes in his notion. But they were fundamental changes, and when we look at Dobzhansky's and Simpson's modifications and extensions of Wright's pictorial representation of evolution, we get a simple, yet succinct and accurate, portrayal of the very essence of modern evolutionary theory.

The first modification of Wright's original conception was the extension of his image of an adaptive field, with many peaks and valleys for all the organisms within a single species, to cover many species at once. Each peak now had its own entire species perched on its upper reaches. The second change was the assertion that it is, after all, natural selection which accomplishes the change from peak to peak. Both were radical departures. Though Wright's own original language itself suggested that differences between species could be explained in terms of the occupation of different peaks, nonetheless his original intent was clear enough: those various combinations of genes, harmonious and inharmonious alike, were combinations represented in individual organisms within a single species. The entire landscape, originally, referred to the genetics of a single species. Later writers were more interested in using an upscale version of the metaphor: each *species* occupied an adaptive peak, and the problem in evolution was to explain how entire species got from peak to peak. The synthesis, as we shall see in a moment, saw the metaphor as an apt one, the kernel of an adaptive explanation for the entire history of life. The difference is crucial: punctuated equilibria offers one line of evidence suggesting that perhaps the normal processes of natural selection (plus random genetic drift) that go on within species may not be appropriately extrapolated as a smooth extension to explain the existence of millions of species, in some 90-odd phyla, occupying the earth for some 3.5 billion years. Yet that is indeed the simple, central contention of the "modern synthesis": what goes on within species, especially natural selection modifying gene frequencies, is really all we need know to explain and understand the history of life.

How does the model work? Simplicity itself—and Darwin comes starkly through. Each species is adapted to a "niche," symbolized by a peak in the adaptive landscape. The peak is a generalized representation, now, not really a bunch of harmonious genes working in concert to yield viable organisms. It has really become all those aspects of the environment to which the organisms of a species are adapted: the food they eat, their predation problems, temperature tolerances and whatnot—all the exigencies of life which an organism faces. And selection keeps the organisms of a species hovering somewhere around the top of that adaptive peak.

That being so, how do we get evolution? The answer seemed obvious to all who considered it, but was perhaps best put by G. G. Simpson, who said (in 1944) that the *real* imagery was more like a choppy sea than a static landscape. In other words, the species tracks the change, at the very worst playing catch-up football with a fickle environment just to stay alive. At best, a changing environment provides the opportunity for a species to invent the truly new, to come up with a new set of adaptations that may, with further environmental change, provide the source for a truly novel array of organisms. Selection, in any case, underlies both the maintenance and the modification of adaptations.

And, it seems clear, environments do change. I have already

described the norm for North America—continental flooding to depths of perhaps 200 feet or more by water of distinctly marine salinity. But more recently, the northern third or so of the continent has been covered by a tremendously thick layer of ice four (some say five) times—all within the past 2 million years. Yes, indeed, environments *do* change. It is, probably, in this case no mere knee-jerk reliance on the Cartesian imagery of a universe in constant motion to think of the inevitability of change in the physical environment. The universe, solar system and the earth are all in motion, and combinations of positions, say, of the tilt of the earth, its wobble and the nature of its orbit around the sun systematically vary over a long term, producing inevitable climatic change on earth over the long term. An equilibrium-minded person with a truly long view could claim that over the millions of years the outer bounds of such climatic change are set, and all change is just long-term oscillation. The evolutionist responds (with some justification, I think) that the adjustments achieved by organisms responding to climatic changes 400 million years ago are in large measure irrelevant when the same climatic conditions reappear in our time—because most of the organisms that made those ancient adjustments are no longer around. There may be a long-term cyclicity to environmental change, but its effect on the history of life is not usefully to be regarded as likewise cyclical: for the most part, the history of life, like the history of the United States, is linear. What lives at any moment reflects the possibilities given what was alive the previous moment and the nature of the differences in the physical world between those moments.

Thus the landscape metaphor, simple as it is, appeals to one's common sense. All it says is that organisms of a species are adapted to the environment (there is design in nature) and that, given the inevitability of environmental change, for the organisms it is literally "change or die." Thus is the standard vision of change within a species smoothly connected to the larger theme of change from species to species, and thence on up, linking up all of life. For trends—increase in brain size within the human lineage, for example—the metaphor simply sees the adaptive peaks shifting in a concerted, gradual fashion, in a single constant direction. It is even possible simply to imagine that environments, hence niches, hence peaks of adaptive "fitness" remain stable for periods of time: a possible explanation for the stability we see in, say, *Homo erectus* and other species occupying a point midway along some directional trend. But that we all concede that environments, in the fullness of time, never remain "the same" is one thing. Concluding that it is inevitable that all species become modified to reflect this change, so that we can render the history of life as a series of interconnected adaptive peaks, ridges and mountain ranges, is another. Punctuated equilibria claims that the histories of nearly all species distinctly emphasize stability—nonchange—over change. Yet Dobzhansky's almost lyrical rendition of the history of life still dominates our thinking:

> The enormous diversity of organisms may be envisaged as correlated with the immense variety of environments and of ecological niches which exist on earth. But the variety of ecological niches is not only immense, it is also discontinuous. One species of insect may feed on, for example, oak leaves, and another species on pine needles; an insect intermediate between oak and pine would probably starve to death. Hence, the living world is not a formless array of randomly combining genes and traits, but a great array of families of related gene combinations, which are clustered on a large but finite number of adaptive peaks. Each living species may be thought of as occupying one of the available peaks in the field of gene combinations. The adaptive valleys are deserted and empty.
>
> Furthermore, the adaptive peaks and valleys are not interspersed at random. "Adjacent" adaptive peaks are arranged in groups, which may be likened to mountain ranges in which the separate pinnacles are divided by relatively shallow notches. Thus, the ecological niche occupied by the species "lion" is relatively much closer to those occupied by tiger, puma, and leopard than to those occupied by wolf, coyote, and jackal. The feline adaptive peaks form a group different from the group of canine "peaks." But the feline, canine, ursine, musteline, and certain other groups of peaks form together the adaptive "range" of carnivores, which is separated by deep adaptive valleys from the "ranges" of rodents, bats, ungulates, primates, and others. In turn, these "ranges" are again members of the adaptive system of mammals, which are ecologically and biologically segregated, as a group, from the adaptive systems of birds, reptiles, etc. The hierarchic nature of the biological classification reflects the objectively ascertainable discontinuity of adaptive niches, in other words the discontinuity of ways

and means by which organisms that inhabit the world derive their livelihood from the environment [Dobzhansky, 1951, pp. 9–10].

What strikes one in this passage is that evolutionary theory—that which explains the entire history of life—really *is* a theory of adaptation. "Macroevolution," all the larger-scale features of life's history, emerges as nothing more than a summation of all those small-scale, adaptive modifications which allegedly are going on constantly within species.

But the general collision remains: we know that environments change, and we have an adaptive model that explains not only the "origin" of new species, but the entire history of life. It is Darwin's vision, buttressed by an understanding of heredity. Yet the data from the fossil record strongly suggest that the metaphor is inappropriate. Before we look at what might be a more accurate metaphor, we should consider a bit further why organisms *do not* change when we all concede that their environments necessarily and inevitably *do* change. For the fundamental mistake in Darwin's vision certainly was not that environments change, or that organisms are adapted to environments, but instead lies in a particular vision of the actual dynamics of nature: What really happens to organisms when environments change?

OF GLACIERS AND BEETLES

There can hardly be a more radical transformation of a woodland landscape than to see it covered with a dense layer of ice hundreds of feet thick. Recognition that there had been several episodes of glacial advance in the not-too-distant geological past was something of an intellectual triumph. The idea simply calls for too extreme a transformation, too much of a switch in the "proper" locales of ice and forest, for the notion to be intuitively obvious. It was a young Swiss, Louis Agassiz, before coming to the United States to teach at Harvard, who saw the similarity between the deposits accumulated around the margins of the montane valley glaciers still very much present in his native country and some older formations that had puzzled Europe's geologists. And as with many another good idea, suddenly many disparate problems seemed simultaneously explained: long stringy lines of poorly sorted silts, sands and boulders (moraines, marking the farthest reaches of glaciers), U-shaped valleys in the mountains and evidence of vastly changed sea levels all answer to a single explanation. Glaciers grow by locking up an increasing percentage of the earth's surficial waters. Ice sheets advance southward, coalescing to form massive continental ice fields, grinding up the landscape before them and distinctly lowering sea level. Twenty/twenty hindsight sees glaciation as obvious, but Agassiz's feat was remarkable, calling as it did for us to imagine such a monumental difference between what we see now and what northern Europe and North America must have looked like only 15,000 years ago. Ten thousand years ago there were still mastodonts—true members of the elephant lineage—wandering around what is now New York City.

There can hardly be a greater amount of environmental change than these extreme oscillations. When and where the ice was present, naturally there were few organisms around at all. In front of the ice sheets, temperature varied from arctic conditions during points of glacial maxima, to conditions roughly comparable to those typical nowadays. In ten thousand years, the New York City region has gone from arctic conditions to a temperate clime. And naturally the animals and plants have changed a good deal too. And the biological changes during the glacial epoch (the Pleistocene) over the past 1.6 million years tell us a lot about what the nature of the biological response to environmental change really is like. For one would expect that adaptive evolutionary change might be accelerated during such periods of environmental stress—along the lines suggested by the adaptive landscape. And another advantage of focusing on events in the close-to-recent geological past is that we can get a handle on the origin of present-day species, much as Gould did with his Pleistocene Bermudian snails. And very much the way a paleontologist in England, G. R. Coope, has done for, of all things, the European Pleistocene beetle fauna.

In a story that is probably apocryphal, J. B. S. Haldane was once supposedly asked what he thought his biological career had taught him about the nature of the Creator. He is said to have reflected a moment and then said: "He has an inordinate fondness for beetles." There was a measure of justice in this remark: There must be nearly a million species of insects alive today, many already described in the scientific literature. Of these, probably 300,000 (some say more) are beetles. Just why beetles are so diverse, so "successful," is anyone's guess, but the fact that they *are* so incredibly diverse just adds fuel to the speculative fire that they above practically all other sorts of organisms ought to show some really rapid rates of change during the glacial waxings and wanings of the Pleistocene. They don't—at least according to Dr. Coope.*

Beetles have a hard integument, and so are readily preserved in fairly good shape in the peats and muds associated with the bogs and lakes of Pleistocene times. Thus the delicate features used by entomologists to distinguish species, such as the detailed anatomy of the internal male genital apparatus, is commonly found in Pleistocene fossils because the integument itself is preserved. Older insect fossils are only rarely that good. In a typical occurrence (foundation excavations in London's Trafalgar Square), a number of readily identifiable species of beetles were discovered "in great profusion . . . where they were associated with bones of straight-tusked elephant, hippopotamus and lion" (Coope, 1979, p. 253). All the beetle species in the deposit are still extant, but none now occur as far north as Great Britain. Thus, in the last "interglacial" (the period of warmth between glacial advances—this last being about 120,000 years ago in England) climes were actually warmer than they are now.

Now, Coope relates many interesting things about Pleistocene and modern beetles in Europe, but the bare facts are these: throughout the Pleistocene, all fossil beetles so far analyzed are identical with modern species. Those from the Miocene (5.7 million years old) are closely similar (some appear actually to be still extant), but for the most part appear to be ancestors or close collateral kin to recent species. But even back in the Miocene, the ecological associations are very similar to modern beetle communities, strongly suggesting that beetle species living nearly 6 million years ago had adaptations very much like those of the species in today's fauna that they resemble so closely.

But there is more. So far the beetle story is but one of a growing list of examples documenting stasis in the fossil record. But Coope's study tells us more, because his beetles come from a backdrop of intense, persistent climatic change. All his Pleistocene beetle spe-

* I am relying here particularly on Coope, 1979.

cies are still alive, and his fossil assemblages are very like recent communities. And the beetle fauna (as well as everything else) in any one region—such as England—kept changing radically as the glaciers came and went. What the picture suggests more than anything else is that simple change in locale is the major organic response to environmental change.

In 1974, in response to an early critique of the original punctuated equilibria paper, Gould and I wrote that there seemed to be *three* possible responses to environmental change: adaptive accommodation, extinction—or simply moving on. As we put it then:

> Surely we "believe" in regimes of natural selection which effect a transformation of the gene frequencies of populations, in a regular and progressive manner. How else can we explain adaptation? We know that selection studies on *Drosophila* in population cages clearly demonstrate the validity of directional selection—as if the work of animal breeders over the centuries hadn't amply done so already. We simply doubt that such regimes could persist over thousands of millions of years, through thousands of generations, without interruption by some more plausible event—such as local extinction or "migration." After all, every time an ecologist or physiologist or systematist examines a population of a species, we are told how thoroughly that population is adapted to its habitat. Are we to believe that a species can exist for a million years gradually improving its adaptation, very happy at the outset and presumably for a long time after that, with such imperfect original adaptation? Isn't it more likely that, faced with a linear change in environment extending through so long a period of time (in itself rather difficult to imagine), a species as a whole will change its area of residence, rather than sit there, grin and bear it, and "adapt"? [Eldredge and Gould, 1974, pp. 305–6].

Coope reached precisely the same set of conclusions with his beetles as we did, and it struck us all that conventional evolutionary theory was overlooking the obvious: of the three possibilities (including adaptive change and extinction), by far the most probable, most common outcome of environmental change was a simple rearrangement of habitats. Whole biotic associations—not simply a "beetle community," but entire ecosystems, such as woodlands with all their associated worms, mammals, fungi, birds, protozoa, other insects and so forth—are simply relocated. And for each species, this indeed amounts to "tracking the environment." But the adaptive landscape sees a species *tracking a changing environment through time*. Coope's beetles confirm what any consideration of the real world should have made obvious to all. Species track *the same environment as it moves around in space*. And if a species cannot track its environment, its usual fate is extinction.

One final example from Coope's beetles reveals how extraordinary the details of resolution of this habitat tracking can be—and how utterly inappropriate the adaptive landscape really is for understanding how nature works. Coope tells the story of *Helophorus aquaticus*, a beetle common in shallow pools in Europe. There are two variant versions of this species; Coope refers to them as "geographic races." One occurs in western Europe, the other in eastern and parts of southern Europe. There is a narrow zone of intermediates where the two races come together. Told apart by details of the male genitalia, fossils show that the two races have been distinct (as distinct as they now are) for at least 120,000 years (which, as Coope points out, for beetles amounts to 120,000 generations). But in any one place (Coope selects Britain), which race occurs depends upon which phase of climatic oscillation one has sampled. So even patterns of within-species variation, in which one race seems more closely adapted to cool conditions than the other (as is the case today—the eastern race is, according to Coope, found at rather high elevations), seem to remain stable. Even *within* species there is reason to question the inevitability of adaptive change given the mere passage of time, and given, too, the documentation of truly profound environmental change. The adaptive landscape fails as a meaningful description of the way that nature is organized and the way evolution really occurs.

But organisms, of course, *are* adapted. It's just that they are conservative. The point, really, for all organisms is to keep going, to continue to extract energy from the surrounding environment—and to keep reproducing should they be successful at getting that energy. Coope's study of beetle distributions' shifting around to match the waxings and wanings of continental ice sheets is a simple story of habitat reconstruction and dislocations. When the vast interior continental sea dried up, *Phacops milleri* apparently became extinct; its rival, its daughter species *P. rana*, was already occupying a presumably very similar niche in the remaining seaway running along the eastern edge of the continent. When the sea once again was restored to the Midwest, *P. rana* came out with it—yet another example of organisms' tracking the shifting distribution of favorable habitats. It seems to be a general rule of ecological life: as long as they can find it, organisms will occupy their accustomed habitat. And since movement of habitats seems also the rule as a response to large-scale shifts in temperature, ice and seaways, we have a built-in explanation for stasis: there is now more than ever good reason to expect organisms *not* to exhibit evolutionary change even in the face of serious environmental modifications.

But this habitat tracking, this keeping pace with life's familiar environments, itself threatens to pose something of an enigma. For in denying that adaptive change is primarily to be seen as a necessary outcome of changing climes, I have argued that it is because organisms *are* sufficiently geared to suit their surroundings that as a rule they do not malleably continue to change to reflect every whim of the environment. This comes close to saying that organisms within a species do not become further adapted because they are already adapted. And at the very least, if this is so, we still need to understand how they came to be adapted in the first place—how adaptations do come to be modified in evolutionary history. If we see adaptation as the central magnet for stability—rather than for continual change—we still need to understand how it occurs.

THE BABY AND THE BATHWATER

There is something of an antiadaptationist backlash going on within evolutionary biology these days. The most common misconception about "punctuated equilibria"—that Gould and I proposed a saltationist model of overnight change supposedly based on sudden mutations with large-scale effects (macromutations *à la* Richard Goldschmidt)—in a way reflects this altered biological mood. For the larger implication of punctuated equilibria is that a simple model of ongoing adaptive change *cannot* be the explanation for the evolution of life as we see it today. So Gould and I are seen as "antiadaptationist": and this despite the fact that we used conventional speciation theory and the notion of adaptive change through natural selection to explain the origin of new reproductive communities (species) and the adaptive modifications of organisms through time! Once again it is the larger patterns of evolution—the quick extrapolation of the elements of stasis and change we see going on within species—blown up to explain the entire history of life, that seem to us to pose the problems for standard evolutionary theory. We never questioned adaptation and selection *per se;* nor, at least speaking for myself, do I see any reason to adopt any but a broadly "neo-Darwinian" position when it comes to explaining how adaptations arise—and, indeed, become modified as time goes by. Rather, it seems to be timing and context over the vistas of geologic time that have yet to be integrated into a thorough understanding of the phenomenon of adaptation.

The antiselectionist, antiadaptationist backlash in evolutionary circles these days comes from a wide variety of disciplines and reflects a rather disparate collection of motivations. A particularly common theme, though, is the frustration that many of us have felt when reading yet another "Just So" story. Kipling's disquisitions on leopard spots, elephant trunks and the wrinkly skins of Indian rhinos archly convey wise pronouncements on silly concatenations of events (and with heavy reliance on the inheritance of acquired characteristics). Creationists would have us read instead the King James version of Genesis I as a simple statement of Fiat: what we see out there in the wild is simply what it pleased God to place there. Evolutionary biology simply affirms that *why* organisms look and act the way they do reflects adaptation through natural selection. Scientists eschewing the metaphorical and supernatural might prefer Kipling's style (I do) and read the Bible as allegory, but perforce must stick to the only naturalistic, materialistic explanation around: natural selection. The problem here, though, is the artlessness of much of the specific treatments that we find the concept of selection has suffered in recent years. Evolutionary biology really began to indulge in adaptive scenarios—scenarios that lacked the infectious charm of Kipling *and* the testability conventionally expected in science—to such an unreasonable degree that a backlash was probably inevitable.

Scenarios about giraffes attaining their long necks and the like are truly a commonplace in biology. The problem is inherent in any theory that seeks to explain large-scale, long-term phenomena in terms of processes that we can observe over the far shorter time scale of days, weeks, months and, at most, a few years. That's the whole *advantage* of natural selection: we really *can* study genetic change in the lab *and* in the wild over brief spans of time. And we can test hypotheses that natural selection underlies change in the lab: we can modify bristle counts on the legs of fruit flies, change the coat colors within lineages of guinea pigs—just as Darwin saw breeders getting higher milk yields from cattle and producing the great variety of domestic pigeons. The assumption (and it seems a perfectly reasonable one) is that artificial selection in the lab and barnyard mirrors a process in nature pretty well. But we cannot test the hypothesis that 50 million years of horse evolution, in which the average size of horses increased, as did the relative length of the face and height of the cheek teeth, and the number of toes diminished, was the product of "adaptation through natural selection." Creationists and skeptical evolutionary biologists alike have seized on the latter point, the former triumphantly concluding that there *is* no "macroevolution," the latter content with the more modest conclusion that selection cannot be invoked to explain the grosser features of life's evolutionary history. Both camps, in my opinion, are wrong. Yet the conventional neo-Darwinian explanation of how it all happens is hardly comforting to those of us who continue to see adaptation through natural selection as a key process in the history of life.

For scenarists of horse evolution, or even the vicissitudes of trilobite families, are prone to paint vivid pictures of the advantages they imagine would accrue to organisms *if* they followed a path of adaptive modification, continually tinkering with the mousetrap, making it bigger and ever more efficient. We have already seen that logic at work in the conventional rendition of "how the human got the large brain." What these arguments really amount to is a plausible rendering of a pastiche of "facts"—notions of evolutionary history typically perceived at a sort of "*Gestalt*" level. In plainer terms, I think the gross aspects of the history of life are habitually rendered in such a fashion to make them *look* like the sorts of events we witness in fruit-fly experiments.

And one might ask why such a distortion of the grosser patterns of the history of life has come about. For it truly seems to me that F. J. Teggart was right all along. The approach to the larger themes in the history of life taken by the modern synthesis continues the theme already painfully apparent to Teggart in 1925: a theory of gradual, progressive, adaptive change so thoroughly rules our minds and imaginations that we have somehow, collectively, turned away from some of the most basic patterns permeating the history of life.

We have a theory that—as punctuated equilibria tells us—is out of phase with the actual patterns of events that typically occur as species' histories unfold. And that discrepancy seems enlarged by a considerable order of magnitude when we compare what we *think* the larger-scale events ought to look like with what we actually find. And it has been the paleontologists—my own breed—who have been most responsible for letting ideas dominate reality: geneticists and population biologists, to whom we owe the modern version of natural selection, can only rely on what paleontologists and systematic biologists tell them about the comings and goings of entire species, and what the large-scale evolutionary patterns really look like.

Yet on the other hand, the certainty so characteristic of evolutionary ranks since the late 1940s, the utter assurance not only that natural selection operates in nature, but that we know precisely how it works, has led paleontologists to keep their own counsel. Ever since Darwin, as philosopher Michael Ruse (1982) has recently said, paleontology has occasionally played the role of the difficult child, stirring up trouble and muddying evolutionary waters. But our usual mien has been bland, and we have proffered a collective tacit acceptance of the story of gradual adaptive change, a story that strengthened and became even more entrenched as the synthesis took hold. We paleontologists have said that the history of life supports that interpretation, all the while really knowing that it does not. And part of the fault for such a bizarre situation must come from a naive understanding of just what adaptation is all about. We'll look in greater detail at some of these larger patterns in the history of life in the next chapter—along with the hypotheses currently offered as explanations. Throughout it all, adaptation shines through as an important theme; there is every reason to hang on to that baby as we toss out the bathwater. But before turning in depth to these themes, we need to take just one more, somewhat closer, look at the actual phenomenon of adaptation itself: what it is and how it occurs.

AT LAST: CHANGE IN EVOLUTION

When Teggart wrote that he saw three themes in human history, with stability dominant, followed by gradual drift in human affairs, and with *real* change the least common element (and typically rapid when it did occur), he claimed to be providing a more accurate rendition of the actual patterns of human history. And he said as much about biological history as well. The picture I have painted of biological evolutionary change so far exactly mirrors what Teggart had to say: when Gould and I wrote the early papers on punctuated equilibria, we were trying to stress stasis, so much so that the language we chose to express our ideas was admittedly deliberately provocative. Yet it *is* true that we never said that some gradual change could not, or did not, ever occur. Just as Teggart cited generation-by-generation modification of local argot as gradual change typical of cultural systems, we have always conceded that *of course* gradual change remains a theoretical possibility—and more to the point, a persistent theme in evolutionary history.

But the point is that such gradual change (I have even seen some in my original *Phacops* example) never seems to *get* anywhere. Through some 5 or 6 million years, there seems to have been a slight increase (gradual and progressive!) in the number of lenses crammed into the eyes of individual *Phacops rana*. The change amounts to a within-species historical "trend." Yet the very next evolutionary event was the evolution of *P. norwoodensis,* an event which saw the reduction of the 17 columns of lenses in *P. rana* to the 15 of *P. norwoodensis*. And, not surprisingly, when the number of columns of lenses was reduced, so too was the total number of lenses in the eye—reversing 6 million years of gradual, progressive evolution. *That* sort of pattern speaks more of "species selection" and related matters (see Chapter 6). Just focusing on the gradual trend within *Phacops rana* for the moment, we see that such protracted change seldom if ever resembles what Teggart called the advent of the "really new." Indeed, most gradual change with which I am familiar in the fossil record seems to be more a to-ing and fro-ing—a sort of oscillation within a spectrum of possible states. Typically a lineage will get larger for a while, then start to get smaller. Indeed, modification in average size of organisms within a lineage remains the most common form of gradual change reported from the fossil record in the paleontological literature.

There are, then, two remarkable aspects to gradual, progressive change in the history of life. The first is something of an irony: geneticists checking the rates of published examples of gradual evolution in the fossil record have found that change typically occurs at an almost unbelievably slow rate, far slower than would be expected at even the weakest intensities of natural selection that geneticists ever consider. In other words, when we paleontologists do produce some examples of protracted, directional and arguably adaptive change through time, change such as the increase in number of lenses in the eyes of *Phacops rana*, geneticists seem to think the rate is too slow to be accounted for by natural selection! There is something amusing in that.

And the other point is really Teggart's: yes, there is some concerted, gradual change, and it even may be susceptible to explanation in adaptive terms. But it doesn't tell us, really, about the advent of the truly new. It never really gets us anywhere. George G. Simpson clearly saw that point in his *Tempo and Mode in Evolution:* at the species level, and at the level of genera as well, Simpson thought the fossil record supported a general gradual interpretation. I, of course, disagree. Yet Simpson saw that large groups—say, the orders of mammals (for example, rodents, elephants, carnivores)—appeared too suddenly in the fossil record to admit of a rational explanation in terms of gradual adaptive modification. Thus Simpson's "quantum evolution," in which he devised means for small populations to leave one peak in the adaptive landscape and climb rapidly up another, thereby suddenly entering new ecological circumstances. In 1944, Simpson was willing to ascribe a good deal of his "quantum evolution" to chance, but by 1953, as far as Simpson was concerned, quantum evolution was just really rapid adaptive change within a lineage, a pattern of quick change that was adaptive throughout. But no matter the details of Simpson's theory: in stressing that the fossil record virtually demanded *some* notion of rapid change, Simpson was reversing the old tendency to *ignore* the facts in favor of the theory. And he was also conceding that gradual evolution just does not get anywhere in producing the truly new—a point Steven Stanley has also been stressing more recently.*

---

* Simpson's "quantum evolution" shares with punctuated equilibria the observation that the fossil record literally requires us to acknowledge periods of rapid evolution. For this reason, it is sometimes said that punctuated equilibria and quantum evolution are really one and the same. They are not. Punctuated equilibria is a theory of speciation and the differential production and survival of species that produces large-scale evolutionary patterns—such as trends. Quantum evolution, in contrast, is a theory of the origin of major new adaptations: it is expressly geared to explain major evolutionary shifts by adaptive transformation within a single lineage. Indeed, Simpson often expressed his belief that speciation is an essentially trivial process, simply because it typically does not involve major adaptive change.

OF NICHES AND OPPORTUNITIES

So why do we have change at all? For that is the ironic twist to current events in evolutionary thinking: we have gotten so far from the "change is inevitable given the mere passage of time" theme that how change actually occurs at all now seems a worthy puzzle. When looking at the evolutionary story of *Phacops rana* and kindred species, we saw that change seemed to accompany speciation. Change seemed really to be more a *function* of speciation, in contrast to the older, Darwinian view, which saw speciation as a function of adaptive change. But in the broader context of adaptive change in general, we need to know a bit more about the circumstances that can, and do, foster adaptive change.

In the world of natural design, be the context a Creator or natural selection, the tradition has always stressed perfection. Indeed, the tendency these days in evolutionary biology still is to develop "optimality" arguments, in which selection is seen as a relentless arbiter that chooses to maximize reproductive potentials in a world absolutely fraught with complexities. What, for example, is the best number of eggs for a pair of breeding birds to fertilize, lay and foster? If the name of the game be to leave as many copies of your genes as possible to the next generation (as the modern theory of natural selection would have it), clearly, then, the pairs would want to produce as many eggs as physiologically possible that will fit in the biggest nest they could build. But it is obvious, too, that too many eggs depletes the hen's own energy resources and, perhaps even more tellingly, crowds the nest, escalating competition among sibs for the food supplied by the overtaxed parents. There is a compromise between quantity of offspring and the quality of their upbringing—a theme not unfamiliar in ruminations on human life. Life is a compromise, and selection, in effect, takes cognizance of these trade-offs between desired ends. Hence "optimality": the "strategy" is to produce the most offspring without sacrificing *their* potential to survive, thrive—and, above all, reproduce.

Such optimality arguments are decidedly more cynical, more realistic, than the "best of all possible worlds" scenarios so common in past evolutionary expositions. For life *is* often a trade-off, and compromise is rife in the realm of adaptations. If it is true that ostriches exhale absolutely no H_2O, nonetheless all other desert-dwelling organisms tested so far perforce do emit traces of water:

their adaptation is less than perfect. Gould has written of the panda's thumb, an *ad hoc* extra, digitlike contrivance fashioned from a wrist bone to allow pandas the better to manipulate their bamboo meals. Gould's basic message here: many organismic adaptations, when viewed a bit more closely, emerge more as gerrymandered, imperfect fits to a less-than-perfect world than as beautiful examples of totally efficient functional design. And the point seems unarguable: we may be amazed at the intricate adaptations of some organisms, but for the most part living creatures are relatively inefficient machines. The energy-procuring behaviors and devices of organisms are better seen as "good enough"—to keep organisms alive and to allow them to reproduce—than as manifestations of ultimately perfect design. Gould has argued, indeed, that this lack of perfection, a sort of *de facto* fall from grace, is not only a more realistic image of the organic world, but also in itself a rather striking imprimature of evolution: less-than-perfect design bespeaks a natural process, based on environmental complexities and the limitations on just how much can be achieved starting with a given species, rather than the design of a perfect system by a supernatural Designer. No watchmaker would try to market such a Rube Goldberg watch as most organisms are more fittingly seen to be.

Ecology offers a parallel problem, up one step from the level of organismic adaptation *per se*. For just as there has been a tendency to see organisms as perfectly adapted to the exigencies of life, there has been a related feeling that all the ecological niches of a habitat are likewise always filled. A niche, really, is the position any local population of a species takes in a local ecosystem—and an ecosystem consists of all the organisms of an area plus all the physical factors that bear on their existence: climate, sunlight and so forth. The operative word is "energy," and local ecosystems have been seen as tightly knit packages of energy flow.

Taking an extreme view, as is common, for example, in the sociopolitical area of "ecological" concern, people commonly argue that nature is a very fine-tuned system of delicate checks and balances. Every species present in a local system has its role, and the existence of the many depends on the continued existence of each component species. Thus every available niche is seen to be occupied—and seen, as well, to be vital to the continued existence of the system itself.

It is no knock on the importance of protecting our truly fragile ecosystems to report that once again the finely honed nature of organic interaction seems to have been a bit exaggerated. Ecosystems do not come tumbling down when one or more of their component species is removed: nature is more robust, more resilient than that. It follows, then, that there must be empty niches out there, as indeed ecologists themselves are beginning to stress. In what open sagebrush rangeland still exists out in the Rocky Mountain states, pronghorn antelope still play, though the buffalo, sadly, no longer roam, at least in anything near their numbers of bygone days.

The point is relevant when we consider adaptive change in nature. For if adaptive change is associated with speciation, and indeed, as I briefly suggested in the last chapter, if adaptive change is largely triggered by speciation, then *if* there are no niches, no new ecological opportunities in which a new species might become comfortably established, then perhaps we have identified one of the controls of adaptive change. Change may appear to be difficult and rare because ecological vacancies are themselves rare. Conversely, perhaps the mere existence of vacant niches in some ecosystems is enough somehow to induce speciation. But while there of course must be a place for newly evolved populations in the economy of nature, the existence of empty niches is a powerful argument against the old saw that nature abhors a vacuum so much that it will create a species to fit the need. Something else beyond the mere presence of ecological opportunity seems to set off adaptive change.

In seeing adaptive change as a process of several stages, we begin, perhaps, to get a bit closer to an understanding of the pulse of such change. Recall that nearly all species are geographically variable, and much of that variation is genetically based, reflecting adaptations to various environmental conditions that occur over a species' entire range. Should reproductive isolation set in, the feeling is that local populations (relatively small—thousands rather than millions of organisms, typically located near the environmental extremes of a species' geographic range) can perhaps take their latent adaptive features and focus them rapidly on the particular set of environmental conditions in which they live. This, as we have seen, is the proposed trigger that releases a rapid phase of adaptive modification. If successful—if that habitat, for one thing, persists—the new species has a chance to survive, and a chance, as well, to last long enough to show up in the fossil record. Thus a picture of haphazard (and random insofar as the ecological opportunities are concerned) adaptive adjustments appears as a function purely of the largely accidental incidents of successful speciation. Organisms within all species are "adapted," and remain so throughout their evolutionary history.

Thus we get a picture of opportunistic speciation. Most species persist for long periods of time, remaining little altered, substantially the same as long as they can recognize their habitat. From time to time new species arise in the lineage, and *if* their special habitat persists, and *if* they are sufficiently ecologically distinct from their close relatives to survive competition (should they ever begin living in the same area), we get a multiplication of not-too-dissimilar species within a single lineage. There should be no inherent direction, in all likelihood, in the pattern of episodic changes from species to species. What we do have at this stage, though, is a tantalizing explanation for "macroevolution"—for those large-scale, grosser patterns in evolutionary history. For if there is no inherent direction in adaptive change between species, and if there is usually no adaptive change slowly accumulating within species, how else are we to explain the patterns of directional change between species—those big trends, like brain-size increase in human evolution—than by some larger-level process of differential control of species origin and extinction: in short, by something we might call *species selection?*

At the end of our 1972 paper (see Appendix), Gould and I approached the problem head on. Though we claimed to be proposing "no 'new' type of selection," what we in effect did suggest was a model of differential success of species over the long run—based on the long-term adaptive superiority of some species over others among their close relatives. We avoided giving the process a name, for one reason because notions of higher-level, or "group," selection were very much in disfavor at that time in evolutionary circles. Not long after, Steven Stanley (1975) addressed the problem and gave the process the eminently reasonable and straightforward sobriquet "species selection." Others in the past had toyed with the explicit notion of species selection, but here, I think really for the first time, there was some real evidence that species are discrete, semi-independent actors in the evolutionary drama. That there was a distinct possibility of some higher-level sorting principle in nature that accounts for particular patterns of comings and goings of species—thereby, among other things, regulating to some extent the adaptive history of life—could no longer be ignored.

And we did not ignore it. In a "where are we now" paper Gould and I published in 1977, we went back to that very notion, embellishing it. Many other voices have addressed the problem since then, and there is by now a rich, interesting and somewhat chaotic mélange of both data and opinions on the subject. Here, at last, we address a topic that is very much on the active agenda of today's evolutionary biology: species selection.

6 PARADOX LOST: SPECIES AT PLAY IN THE EVOLUTIONARY GAME

FRUSTRATING OCCUPATION, trying to think about fossils in evolutionary terms. On the one hand we have a theory deeply rooted in genetics, and on the other the scattered and partial remains of thoroughly dead organisms. The patterns of stasis and change that do come through this filter of death and destruction really are so gross that even if we miraculously knew the genetic basis of the features we see in some of our fossil specimens, there would still be the horrendous difficulty of specifying just what happened to those genes during the eons that elapsed between our imaginary "genetical" fossil samples. The record is too coarse to be of much use to a geneticist, who requires for most purposes generation-by-generation monitoring of gene frequencies.

How absolutely intriguing, then, that out of a reconsideration of basic patterns of stasis and change in the anatomies of fossilized organisms—a reappraisal of the basic evolutionary message of life's history that led directly to punctuated equilibria—we receive, almost as an afterthought, an accidental by-product, the kernel of an idea that promises to relocate paleontology smack back into the middle of the game of evolutionary theory. For if it is true that species really are like individuals, if it is true that they come and go at least partially independently of one another, then obviously any tinkering with the rates of "births" and "deaths" of some species *vis-à-vis* others will introduce an element of bias into what species will occupy the earth in the future. Natural selection provides the bias in the passage of genes from one generation to the next, a simple function of the varied reproductive success of members of the parental generation. But if species are coming and going, and patterns—such as the episodic trend toward increase in brain size within our own lineage—suggest that there is some regulation of which species occurs where and for how long, then we have the basic ingredients of some exciting new theory. We have the potential for a form of "selection" above the level of genetic properties of organisms-within-populations. We might, in fact, be seeing a process of "species selection." And paleontologists, for a welcome change, are the very ones who have the data needed to investigate such notions: at last, the very coarseness of our data is just what is needed for us to see the forest for all the trees. Species stick out like sore thumbs in the fossil record.

BACK TO GONDWANA: ONE FAMILY'S STORY

There is a minor, yet persistent, theme throughout biology's history which tends to see the creatures of the southern hemisphere as descendants of earlier northern forebears. Reminiscent of the "Christopher Columbus effect" (which tends to see most "New World" creatures as scions of Old World ancestors—as indeed I claimed is true of *Phacops rana!*), this northern-hemisphere myopia seems innocently enough come by: most of the early students of the animals, plants and fossils of South America, Africa and Australia were Europeans who compared what they saw with what they knew about the world on the basis of intimate experience back home. Much of the time, these old naturalists took what they found in these "austral" regions as variant versions of northern, "boreal" life forms. And that is really not a bad way of looking at the southern-hemisphere biota, past and present: often similar, analogous, but by no means identical to northern forms. But it is another thing, of course, to conclude thereby that the southern creatures are derived versions of the northern forms—any more than it would be sensible to claim that northern forms invariably spring from a southern source. As is usual in such matters, the situation is much more complex than any simpleminded rule of thumb would have it. For the most part, wildlife of the southern hemisphere is best understood as a world apart, something that developed on its own.

John M. Clarke, State Paleontologist of New York at the turn of the century, was one of the best of the lot of second-generation U.S. paleontologists. It was he who really gave us the fine early work on *Phacops rana* (in a monograph, written with the grand old man of American paleontology, James Hall, in which Clarke pinned down the European affinities of *P. rana*) and the paper on eye anatomy in 1889 as well (see Chapter 3). Clarke had an inordinate love for the Devonian—particularly, it seems, for the Lower Devonian; nearly every summer for several decades he made the laborious trip overland, and thence by steamer, to gorgeous Gaspé, near the mouth of the St. Lawrence River in Quebec Province. There he would haunt Lower Devonian locales, by far the most spectacular being Percé Rock, which Clarke reckoned contained more than 60 *million* specimens of just one of the dozen or so trilobite species buried within. It was only natural that Clarke would get involved in some of the earlier work on the Devonian fossils beginning to turn up in respectable numbers from South America.

In 1895, Clarke published (in side-by-side English and Portuguese texts) "Fosseios Devonianos do Paraná"—and, some years later, illustrated additional Brazilian fossils together with a small, yet distinctive, collection of Lower Devonian trilobites from the Falkland (or Malvín, depending upon who owns them) Islands. To this date, Clarke's has been the last word on the Falkland fossils. At first indeed seeing the dominant sort of trilobite from these southerly reaches as but a variant version of forms familiar in our own northern strata, Clarke later in his career began to realize that the fossils of the Falklands, while different in particular details from those of southern Brazil, nonetheless bore a sort of general (or familial) resemblance to them, and to specimens collected in the Amazon basin, from Bolivia, high in the Andes—and to specimens known as well from southern Africa. In an absolutely typical microcosm of the general groping for understanding that has come from northern minds confronting southern problems, we are only now beginning to understand how aloof, how biotically independent, the northern and southern hemispheres have been from each other for a good 20 million years in the Lower Devonian. And during this era of isolation, a spectacular flowering of a single trilobite family—containing alike some very conventional-looking species and some of the oddest trilobites that have ever lived—arose, proliferated, flourished and disappeared. All were descended from a single common ancestral species, to judge from the retention of some unique features that all members of the family, no matter how bizarre, wear as a stamp of their pedigree. Situations such as this, where sampling is good, the ages of the samples are at least fairly well grasped and the evolutionary history of the group can be worked out in detail, are the very meat of evolutionary analysis in paleontology these days. And the story of that family boils down to the comings and goings of different species within various subdivisions of the family—superimposed over a pattern of adaptive differentiation which, just as in the case of *Phacops rana*, must have been under the control of natural selection.

The world, remember, was beginning to coalesce up north as what we now call North America was moving north and east to collide with westwardly drifting continental Europe in Lower Devonian times. But the supercontinent Gondwana in the south had been assembled from time immemorial—if "assembled" is even the right word for it. Gondwana seems to have been around for hundreds of millions of years, with no clearly discernible pattern of coming-together-then-breaking-up, as was the case in the north. Eventually, toward the end of the Paleozoic, *all* the continents came together to form Pangea—which, among other things, must have put a bit of an extra wobble in the earth's spin, what with all that uneven disposition of continents and oceans over its surface. (Twenty years or so ago, Australian geologist S. Warren Carey pointed out that *today's* distribution of continental masses over the earth's surface resembles a tetrahedron—a respectably stable configuration of continents on our spinning globe.)

But not all past distributions of plants and animals fall neatly out of the hopper of continental drift. The portion of Gondwana we call Africa today was not overly far from Europe, and thus from North America, by Middle Devonian times. And the *Phacops rana* stock is very much in evidence in northernmost Africa. Disposition of seaways, and especially of current patterns that determine just where larval trilobites, brachiopods, clams and snails will float, are vitally important too. The unique austral Gondwana trilobite fauna is simply not found, for example, in Australia—whose Lower Devonian trilobite fauna looks to me virtually identical to species we get in Lower Devonian limestones of the Appalachians. A few years ago fellow trilobitologist Allen Ormiston and I put together a rough-guess map of the distribution of land and sea in the southern hemisphere during Devonian times. We estimated, too, the circulation patterns that seemed to make sense on general oceanographic principles, currents that would help us understand why Australia looks so "northern" while its Gondwana cousin continents look so appropriately southern.

Easier to fathom are some of the physical conditions of those ancient seas. Unlike the warm shallow seas crowded with corals and other earmarks of the tropics that we saw in Middle Devonian North American waters, the trilobites, brachiopods, clams and snails of this special southern province were living in truly frigid, arctic conditions. The deduction comes not so much from the organisms themselves, but mostly from paleomagnetic studies, which place the south pole only a few hundred miles north of Cape Town, and from what is *not* there in the fauna: corals are virtually absent, as indeed are any truly biogenic (that is, organism-created) limy sediments. Most modern corals are confined to a zone between 60 degrees north and 60 degrees south latitude. (Those which form colonies and build reefs are restricted to a much narrower zone centered on the equator.) And paleomagnetic evidence suggests that the farthest north *this* fauna got was 60 degrees *south* latitude. A polar fauna indeed.

And the trilobites do bear out this inference. Just as many of the largest, spiniest living arthropods, such as the Alaskan king crab, or the largest of the American lobsters, typically prefer very cold oceanic waters, some (but by no means all) of the Lower Devonian Gondwana trilobites grew to inordinate sizes, and some were also unusually spiny. Three or four species grew to well over a foot in length, though none of these belong to our star object of macroevolutionary attention—the remarkably fecund family Calmoniidae.

Calmonia itself, a modest little trilobite with a few poorly known species from southern Brazil and the Falklands, has lent its name to the entire family. And in a way this is fitting, as *Calmonia* fairly does typify the family. The central region of the head of all calmoniids is their hallmark—together with their unusually small eyes—and *Calmonia,* thankfully, is no exception. It is the peculiar excrescences around the front and sides of the head, along the body and around the tail that mark most of the evolutionary activity in this family—permutations and combinations of spines that seem to present an endless variety of species. What these spines are *for,* though, is anybody's guess. Trilobites, the oldest major division of the ar-

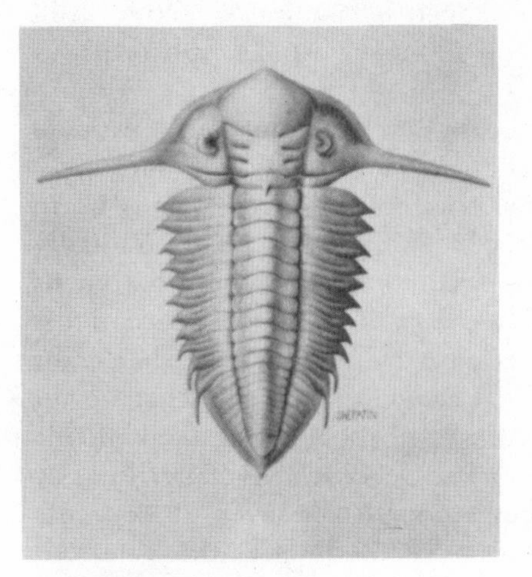

Deltacephalaspis,
a Bolivian calmoniid trilobite.

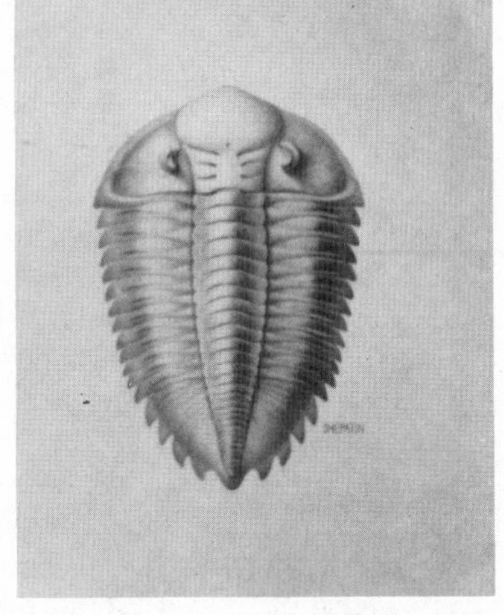

Metacryphaeus, a calmoniid
found throughout Devonian Gondwana.

thropods, became extinct 250 million years ago, and living crustaceans and horseshoe crabs provide insight but no sure guideline to the finer points of trilobite behavioral anatomy. The diversity in spine configuration could conceivably reflect differences in feeding behavior, or swimming abilities—or, as I think perhaps most likely, provide a means of recognizing prospective mates within the same species.

But there really is no way to tell. And that has always been a near-fatal barricade to evolutionary studies based in the fossil record. *If* all there is to evolution is adaptive change, it follows that the analysis of how anatomical structures are actually put to use for the good of the organism would be a paleontologist's natural entry into the world of evolutionary analysis. And indeed, the technical literature is littered with countless examples of imaginative analyses of "functional morphology"—that is, informed speculations on why those trilobites had those spines, elephants their trunks and so forth. For to know how these sundry items were used is to know how they got there: they are adaptations shaped by natural selection. And adaptation is all there is to evolution—or so the subconscious thought processes run.

But there are some creatures, such as this remarkably variable array of southern Devonian trilobites, which simply defy such analysis. This is *not* to say one must always throw one's hands up in abject resignation: in a remarkable study, Dan Fisher virtually proved that a Jurassic horseshoe crab *must* have swum flat on its back at a speed of about 20 centimeters per second. He turned this trick by analyzing the physical reasons why modern horseshoe crabs swim on their backs but at a slight angle, and at somewhat slower speeds. He then made models of the fossil species, put them into his tanks and discovered the speed and angle which the flatter Jurassic species must have utilized if they too were to enjoy a "frictionless backstroke" as they used their legs to paddle through the water. All Fisher assumed was that Jurassic horseshoe crabs would have had such a frictionless backstroke—which seems eminently fair and reasonable. Functional morphology, as this business is usually called, really *can* be done sometimes with fossils.

But Fisher had living horseshoe crabs, and that happenstance allowed him to understand the basis of how they swim given the shapes they have. His task was then to explain why two horseshoe crabs differed in some respects. I haven't any really solid idea to start with when it comes to divining the function, thus adaptive significance and evolutionary origin, of these spiny Gondwana trilobites. And in ordinary evolutionary contexts, that's the kiss of death. Just describe them and go on to some other problem.

But there is another way. Let us suppose that all those spines and odd-shaped heads and tails *are* adaptive. Let us suppose they represent the reorganization of ancient anatomies by natural selection. We take all that for granted—a casual way of saying that we accept it axiomatically. And this seems only right: after all, natural selection as a testable scientific proposition requires intimate knowledge of the genetic basis of anatomical features *and* an accurate charting of change as each generation rolls on by. Strike three for the fossil record. But if we simply treat adaptation and natural selection as a "black box," can we go on and find out anything else about the nature of the evolutionary process by looking at this family's history? I think we can—and here is the major legacy of punctuated equilibria.

There are at least 75 calmoniid species so far known. There were doubtless more: some of those 75, still not officially named or described in the scientific literature, are represented by a few specimens sitting in museum cabinets. But there are enough known by now to give us some obvious features of family history. For starters, not all species were present throughout the 25–30-million-year interval that marks the total life span of the family. The most complete temporal records appear to be in the Andes, particularly in Bolivia, and here there are at least four distinct, definable phases to this evolutionary drama.

The first phase is modesty itself: it consists of but two species, both belonging to the primitive ancestral genus *Andinacaste*. Acastid trilobites were worldwide in the preceding Silurian Period, and *Andinacaste* is but an acastid with exceptionally small eyes. It was this little offshoot of cosmopolitan acastids which apparently lay at, or certainly near, the roots of the entire family.

The second phase, also best known from the Andes, saw a modest proliferation into some 10 genera. Some retained a rather primitive leer, but others were among the more bizarre of all—including the egregious *Deltacephalaspis*, with one of the most outlandish heads ever to grace a trilobite. There then followed a long interval, a third phase that fashioned the vast bulk of Lower Devonian calmoniid history, in which a succession of species sporadically replaced one another both in time and over space, with different species appearing in different subregions of this austral realm. For the most part, the replacement species seem to have been just variant versions within the same well-entrenched genera. Little major change, in other words, accumulated, though there were some novelties and trends along the way. For the most part, we see a gross sort of stability during this third stage, in which the ecological assemblages of these trilobites and other fossils develop a sort of predictable sameness. In the fourth and final phase, we see fewer species, and the beginnings of biological connections with the rest of the world—all a last gasp as the continental seaways began to give way to terrestrial environments. The end came as the seas finally yielded to land, and through no failure of these creatures in any cosmic sense other than their pure inability to find their cold-and-shallow-water marine habitat any longer.

But such a rapid-fire and coarse summary of family history misses all the fine points. Phase 3, for example, in which many genera settled in for a long period of business-as-usual stability, losing a species here, gaining a rather similar new one there, with the invention of the truly new for the time being placed on hold, looks to be a simple entrenchment, a consolidation of the earlier, more experimental phase. It wasn't. Virtually none of the trilobites of that initial burst of diversification ("Phase 2") left any descendant species. The vast majority of species in that long third phase were derived from but one or two species in the initial burst. But before we become immersed in many more of the details of this intricate history, it would be as well to ask: Just what are we looking for? What are the questions? For it surely is the world of ideas that tells us what to look for, and why we might conceivably be interested in the arcane intricacies of species' comings and goings in a long-dead group of trilobites.

SPECIES SELECTION

It seems obvious to every paleontologist that all species have a finite life expectancy. It is only slightly less obvious that some lineages characteristically produce more species—show higher rates of speciation—than others. And some species seem more prone to extinction than others. The result: a systematic bias in the world's species composition from interval to interval in geologic time. Since we attribute both stasis and change in gene frequencies to a bias in the births and deaths of organisms with varying genetic propensities toward "natural selection," it seems only proper to dub this bias "species selection." Whatever we call it, if there is a nonrandom element to change in the species composition of the living world, we have identified an evolutionary mechanism overlooked in most conventional theories of evolution.

So we must ask if there is anything intrinsic to species that would make them characteristically prone to extinction—or conversely, enable them to stand a better chance *against* extinction. And we would also like to know whether those same properties might also convey a propensity toward speciation—or, again alternatively, act as a sort of damper on the tendency to bud off descendant species.

The answer, in general, to these questions is "yes," though I cannot pretend that the form *my* answer takes is the only correct solution—or even, in the long run, will prove to be correct at all. And in a book about punctuated equilibria, it is important to emphasize that my own views on macroevolution—what it is and how it happens—are somewhat different from Gould's. Some of his post-1977 speculations have indeed approached macroevolution primarily as a problem of the modification of adaptive structure, *per se;* I have preferred to focus not on the transformation of organismal characteristics (which requires data and theory from genetics and developmental biology), focusing instead on features of organisms and species that may bias the rates of birth and death of species. I persist in seeing macroevolution as fluctuations in numbers of species in a lineage, with concomitant implications for the amount and kind of adaptive change that will accrue. However incomplete the work may now seem, at least it is a start, and at the *very* least it brings in some long-known but little-appreciated features of life that really do seem to have a bearing after all on the evolutionary process.

It is fashionable these days to distinguish "pattern" from "process"—patterns being the result of process, the actual historical events that have occurred, for example, in evolutionary history. Yet, as Thomas Henry Huxley pointed out in the last century, many of the "processes" we think we see in nature are necessarily just cases that remind us of other, very similar phenomena in other organisms. Thus the ecological "process" that seems to me to be so important in biasing the births and deaths of entire species at the very least springs from some very well known, and almost anecdotal, truths about the nature of a wide variety of organisms.

Around the turn of the century, a paleontologist at Cornell, one Henry Shaler Williams, observed that there were really two sorts of fossil species. Williams' dichotomy stands out most clearly when closely related species are compared. His own specialty was in Middle and Upper Devonian fossils, among them spiriferid brachiopods, which have a distinctive shape that gives them the look of airline captain's wings. In any case, Williams thought he saw two different syndromes in his brachiopod species. Some seemed to be restricted to very limited areas—say, the near-shore areas of the continental seas, confined to what is now the Catskill Mountains in east-central New York. And there was often more to their limits than just *space:* such species typically occurred in a very narrow range of physical environments, as far as the kinds of muddy bottoms and varieties of other organisms with which they occurred could reveal. And such species tended to have relatively brief total time spans. (This has nothing to do with *stasis*, or relative nonchange, throughout a species' duration; it is simply a matter of how long, on average, species will tend to persist. Some live longer than others.)

But that wasn't all. Williams studied the individual fossilized organisms of each of his species very carefully. And he was able to reach two further generalizations about those species which seemed so limited in their distributions in space, time and habitat. The organisms belonging to such restricted species were all very similar; there was simply less variability in such species than is often observed in other species. And they seemed highly distinctive, frequently sporting some anatomical peculiarity or other—say, an exaggerated set of ornamental features on the outside of the shell—that made those shells stand out as distinctive from the run-of-the-mill generalized sorts of spiriferids more typical of that group of brachiopods.

And the correlations seemed to work in reverse just as faithfully: the more generalized sorts of spiriferids seemed in general fairly variable. They typically occurred over far broader areas, in a greater diversity of habitats. And they seemed to hang on for relatively greater chunks of geological time.

The basics of the story are clear enough: creatures able to exploit a greater range of habitats will naturally tend to be more far-flung, merely reflecting their greater ecological opportunities. And they will have a better chance of continuing to track those habitats as time and environments change on them, so they will persist, on average, longer. Their greater variability merely reflects the different characteristics they must develop simply to fit into such a comparatively broad array of habitats. And that is how Williams left the matter: an interesting sort of pattern he felt had significance far beyond his brachiopod examples. And indeed he was right: the pattern explains much about the history of life.

But why were those far-flung, variable species so drab, so singularly lacking in the sorts of anatomical filigree so typical of that other great class of species? For the distribution of a species in habitats, and through time and space, might seem to be a "property" of a species, but it is really the sum total of when and where all those organisms in any species were living. Even variability, based as it is on examination of individual organisms, is a sort of group-wide summation of the organisms themselves: Williams' individual brachiopod specimens were not variable. It was just that in some species, all specimens seemed more or less the same, while in other species the individual organisms differed somewhat among themselves. But when we consider anatomical specializations, we get down to observable properties of individual organisms. Every specimen of *Deltacephalaspis magister* from the Lower Devonian of Bolivia had that whacking great set of lateral horns on its head.

Anatomical peculiarities are actually a further clue to what seems to underlie Williams' two types of species. Suppose these anatomical peculiarities reflect some specialization, a specific adaptation to aspects of an organism's environmental circumstances. The hovering wings of hummingbirds, and their long, curved nectar-sipping beaks and tongues, reflect such a specialization. But the distinction can be even more subtle than that: some hummingbirds cruise broad ranges and sip from a wide variety of flowers, while others restrict their nectarizing to but a few plant species. In short, some organisms seem to be focused very narrowly, specializing on only one or two energy sources, or tolerant to a very narrow range of temperatures or salinities. Others are more loosely integrated with the rest of the world—flexible, able to live as happily on worms as on clams and to withstand huge fluctuations in temperature, salinity and evey oxygen availability—as in our living horseshoe crabs on the east coast of North America. But such variation in how organisms perceive, and behave in, their environment is not distributed randomly: all organisms within a species behave more or less alike. That is why we say some species are "broad-niched," or "ecologically generalized," while others are "narrow-niched," or "specialized." But what we mean when we say such things is that the *organisms* within those species are generalists or specialists.

So we have come full circle: unexpectedly, we are once again staring at adaptation. But now adaptation seems more a key than a blind alley in the endeavor to understand something about evolutionary dynamics. In the old metaphor of the "adaptive landscape," it was always assumed that a species would concentrate near the top of the peak, which represented some acme of adaptive perfection. It seemed inevitable that selection would always be at work perfecting a species' collective adaptation. But this perfectly understandable expectation was utterly confounded with an assumption of *speciali-*

zation: selection, in its drive to perfect adaptations, should be eliminating variation.*

But let's look at the problem in a slightly different way. Suppose that there is more than one style, one "strategy," to exploiting an ecological niche. In fact there seems to be an entire spectrum of "widths" to ecological niches. Some are broad and capacious—such as the niche we find modern horseshoe crabs seemingly happily exploiting. Others are indeed quite narrow, focused sharply on a limited range of resources, like some of the species of African antelope that feed on only one or two species of grass. The adaptive landscape carried the unspoken assumption that the epitome of niche exploitation was "perfection" of adaptation through specialization. But if a species is broad-niched to begin with, continued selection, fine-tuning that "adaptation," will tend to favor the physiologically tolerant, not those aberrant organisms which stray toward specialization. Normal horseshoe crabs do indeed dine on a wide variety of marine worms, clams and snails—virtually whatever they can get their peculiar jaws around. Any young horseshoe crab that is particularly fond of worms, or, more realistically, fails to become strong enough to crack open clamshells, has less than the full range of resources normally available at its disposal. It is consequently less likely to succeed at the game of life, with obvious consequences to its probable success in leaving genes to the next generation.

Ecologists often speak of "habitat perception." Some organisms seem to "perceive" their environment as very coarse-grained: they are the generalists. Other species see the environment as fine-grained: they are the narrowly focused specialists. *All* species are well adapted, and we can safely assume that selection will keep them there. What does this have to do with speciation and extinction?

Again, the connection seems clearest when we follow Williams' lead and compare two closely related groups of species within the same lineage. For example, *Turritella* is a very common genus of modern marine snail. The individual snail lives by wedging its tall spire into the bottom muds and filtering food particles directly from the seawater—altogether a somewhat unusual mode of life for a

* Yet variation was clearly important for future evolution in the species—a conundrum, that has bedeviled orthodox theory for years, and one that I have examined at length in my book *Unfinished Synthesis*.

snail. Paleontologist Judith Spiller studied all *Turritella* species that have lived along the Atlantic coastal plain over the past 18 million years. She came up with a pattern strikingly like the one that Williams reported so long ago for his Devonian brachiopods.

There are, in short, two kinds of *Turritella* species. Some species are variable, occur up and down the coast and range through large chunks of time. *Turritella variabilis* (the specific name simply means "variable") is your garden-variety *Turritella*, ornamented rather modestly with just a few spiral lines. The other sort of *Turritella* is something else again: the species all seem to be restricted to semi-isolated portions of the Atlantic coast, and show typically very short durations in time. They are highly ornamented (for a snail, at least), and vary very little. But more is at stake here than just another example like Williams' brachiopods. For the two sorts of *Turritella* belong to two closely related, but distinct natural lineages. The variable species form one coherent ancestral–descendant sequence, as do the seemingly specialized species. And perception of this segregation into natural groups made one further statistic take on added significance: the specialized species outnumber the apparent generalists by more than 3 to 1. In the last 18 million years, there were 13 species in the highly ornamented lineage as against but 4 in the *variabilis* group. The plot thickens: in the *variabilis* group, both the appearance of new species and the disappearance of old went on much more slowly than in the related lineage. But the rates weren't really at a snail's pace—for that is the whole point: we are comparing two subdivisions of the same basic lineage of gastropods.

What's more, Spiller found that living relatives of the two lineages differ in their ecological requirements, and especially in the critical factor of reproduction. The generalist, *variabilis* types send hordes of larvae up into the water column, there to be spread widely in each generation. The larvae take a long time to settle and metamorphose into adults. The other sort of *Turritella* produces larvae that metamorphose quickly into adults—hence spend little time floating around in the currents. And a very convenient feature for the study of snails is that they retain the larval shells, perched as a minute coil atop the adult shell, throughout life. Any snail shell, fossil or recent, provided it is well enough preserved, will tell you something about its larval longevity. With their dispersal ability so dramatically restricted, Spiller saw that the specialist lineage of *Turitella* would produce an array of species each time sea level rose and the Atlantic coastal plain became divided into a series of embayments—each one with its own distinctive species of the highly ornamented lineage. Meanwhile, the same tramp generalist species would continue to ply its way up and down the coastline, seemingly unaffected by the vagaries of oscillating sea levels and shifting shorelines.

Williams and many others before and after him always felt there was a connection between a species' adaptation and its chances for long-term survival: in modern parlance, broad-niched species ought to be able to stave off extinction longer than narrow-niched species—if for no other reason than that generalists are usually farther-flung and engaged in exploiting a greater spectrum of habitats, enhancing their chances of tracking their habitats when their world is disrupted. Now, studies like Spiller's were indicating that niche width also had some real implications for speciation, the *births* of species as well as their deaths. There are several ways in which this might work—and herein lies some more fresh dispute in evolutionary circles these days.

One way to imagine the connection involves another ecological notion already familiar from an earlier, different context: competition. Acknowledging that at present it is trendy in some ecological circles to dismiss competition as illusory, there *is* the matter that much of the biological realm looks as if it were organized to *avoid* competition. Similar observations show the relevance, I think, of competition to understanding why specialist lineages typically have many times more species than their generalist kin. When we count up numbers of closely related species in an area—preferably species of the same genus, such as all the species of *Turritella*—we get yet another highly repetitive pattern. Specialized species tend, very often, to occur together, while closely related generalists rarely do. The supposition is that broadly niched species are more prone to niche overlap, an unstable, competitive situation in traditional ecological theory. It boils down to a habitat's not being big enough for the two of them.

The reason that a habitat can support more closely related specialist species than generalists seems clear enough: it is a matter of dividing up the ecological pie—meaning, for the most part, energy resources. All animals, for example, require some form of external food supply to obtain the energy *and* matter (such as amino acids) vital to continued existence. Specialists, each focusing on one particular food item, can easily coexist: the great proliferation of antelope on the African plains reflects this phenomenon to a great extent, as the species living cheek by jowl in general do tend to partition the various grasses and bushes that provide the fodder for these vegetarians. And in the lakes of the east-African rift-valley system, the great species flocks of cichlid fishes mirror the very same phenomenon. In several of these great lakes there are two groups of cichlid genera —the generalists (*Tilapia*) and the specialists (*Haplochromis* and close relatives). Each *Haplochromis* species has a unique mode of feeding, and there is little or no demonstrable conflict over energy resources. In contrast, the closely related but more primitive genus *Tilapia* is generally represented by but one or two species—and, sure enough, those species are ecological generalists, spread throughout the lake in great numbers, and willing and able to dine on a wide variety of food items.

But it is hard to imagine how division of the ecological pie by specialization will itself induce new species to appear. A simpler picture, I think, is that in lineages whose species are already rather narrow-niched, new species become reproductively isolated in the usual way—but they have a slightly better chance of surviving because of their propensity to specialize. Reproductive isolation without a further narrowing of adaptation will lead to survival only if the parent and daughter species forever live in different areas. Should they ever come into contact again, the ensuing contact would drive one or the other out. But should the daughter attain some degree of specialization itself, it will be able to occupy the same general habitat, but in a recognizably distinct niche. In this way, the number of closely related species in one area may grow by simple accumulation.

The flip side of this idea is that generalist species bud off descendants just as rapidly as specialists do, but the probability of survival for the fledglings is lower because broader-niched species simply have greater difficulties in sharing resources. Some botanists have painted a picture of new species of grasses appearing yearly along California's roadsides, species reproductively isolated from their widespread parental species, in this case, by sudden mutational change. But the new species typically remain nearly identical ecologically to the parental species. Thus small stands of the new species have little hope of survival, swamped as they almost always are by their vastly more populous, well-established ancestor.

Such imagery is appealing: it says that generalists and specialists have equal probabilities of speciation—creating new descendant reproductive communities—but the survival rate of fledgling specialist species is greater, giving them the appearance of faster rates of speciation. But there is another way the system might work, a suggestion made by paleontologist Elisabeth Vrba. She notes that species which are ecologically generalized usually also display a very general pattern of reproductive behavior, while ecological specialists tend to have rather precise behavior patterns and signals for reproduction. Concluding that specialized reproductive systems are simply more easily disrupted, Vrba sees a faster intrinsic rate of speciation in specialists—rather a different way of translating the generalist/specialist dichotomy into differential rates of speciation.

Nor should we assume that ecological strategies are the only causes of different rates of species appearances. No matter which, if either, of the two theories invoking niche width is the correct interpretation, there may be other controls over species birth and death rates—and indeed, there are other suggested mechanisms in the literature. But there is a goodly measure of evidence that certainly does implicate niche width in apparent rates of speciation, affording us a pretty good start in determining just what the mechanisms might be that bias the appearance of new species in some lineages over their appearance in others.

The other side of the coin, as I have already mentioned, is less difficult to see: it has long been felt that those species which do not put all their adaptive eggs in a single basket stand a better chance of surviving the vicissitudes of ecological reality. The geological time scale of the past 600 million years is divided up into a series of finer and finer units, each marked by characteristic suites of fossils. The larger units reflect the more radical changes in the very composition of life. And nearly every burst of evolutionary activity represents a rebound following a devastating episode of extinction. The truly severe extinctions took out up to 90 percent of all species then on the face of the earth. But there is also "background extinction"—sort of a normal ticking away, in which species drop by the wayside unaccompanied by a whole series of ecological confederates. And it is this sort of quiet ticking away of the extinction clock that seems most relevant to differential species survival: Most of us, whenever the word "selection" comes up, are still accustomed automatically to think of the deaths of some and survival of others, forgetting that biases in births can be equally important. Indeed, as we have seen, geneticists have gone nearly wholeheartedly the other way, defining natural selection as differential success in producing new organisms, overlooking their deaths almost entirely.

So too with species selection: nearly all the early discussions of "species selection" in the decade following the publication of our paper in 1972 focused on the relative *survival* success of species—carrying the implication that somehow one species was "superior" to another. But if survival usually goes to the generalists, and a broad-niched ecological strategy is not inherently superior to a narrow-niched adaptation (the two being merely different ways of dealing with the environment, equally good insofar as the *organisms* are concerned), the relatively longer life that generalized *species* seem to "enjoy" is merely a side effect of the physiologies of organisms.

And there really is no doubt that generalized species on average last longer than their more specialized close relatives—just as H. S. Williams said so long ago. Perhaps the best example of the entire phenomenon comes from E. S. Vrba's (1980) study of African antelopes. In contrast to my Devonian trilobites, antelopes are still very much alive, and their ecological behavior has been intensively studied for a number of years now. And antelopes also proffer one tremendous advantage: mate recognition is known to be accomplished through a number of physical and behavioral patterns—not least the size, shape and mode of twisting of the horns. Each living antelope species sports a distinctive horn style—often in both sexes, sometimes only in the male. This means that we can use their horns to tell the species apart—not critical, perhaps, with living animals, as most antelope species are readily distinguishable on a host of features. But it makes a big difference with fossils: portions of skulls with at least the bases of the horns intact are reasonably common in deposits in eastern and southern Africa, providing the key to Vrba's analysis of antelope evolution.

And sure enough, once again we find old H. S. Williams' pattern. Vrba's best-analyzed example involves two antelope tribes with the forbidding names "Aepycerotini" and "Alcelaphini." But the animals themselves are reasonably familiar: Aepycerotini turn out to be impalas, while the Alcelaphini include the two living species of wildebeests (gnus) as their most familiar members. Less well known (at least in North America) but equally eye-catching bonteboks, blesboks, tsessebes and hartebeests are also included along with the wildebeests—all in all, 7 living species of alcelaphines.

Wildebeests have long struck biologists as perhaps the most "highly evolved" of Africa's antelopes. A loaded expression, "highly evolved" just usually means that gnus have become rather highly modified from the presumptive primitive antelope condition. And with their elongated snouts and high shoulders, wildebeests really do seem a far cry from, say, a Thompson's gazelle. Or, for that matter, from impalas, which to untutored eyes look like perfectly ordinary antelope. Yet Vrba found that impalas really are the most closely related of all antelopes to the highly distinctive group of alcelaphines—the wildebeest cadre. The two groups are phylogenetic "sisters," and impalas seem to be close in general appearance to that entire corner of the antelope realm. Impalas simply haven't changed much since the entire group of antelopes first appeared.

Delving back into the fossil record, with fossilized horns providing a sure guide, Vrba found that the impala-wildebeest group arose some 5 million years ago. And she found more elements of what really must be called "Williams' Pattern": there is but one species of impala alive today, and it occurs over pretty much of the length, if not the entire breadth, of the African continent. The fossil record documents the continuous presence of impalas from the Miocene on —yet Vrba found at most two, and most likely only one, other species of impala in the fossil record. Impalas really haven't changed much at all since their evolutionary invention; and they have always occurred, it seems, as single, far-flung and long-ranging stable species.

Not so with the sister Alcelaphini. To date Vrba has documented some 25 fossil alcelaphines to accompany the 7 extant species. Each species has a geographic range much narrower than that of impalas (though the long migratory trek of modern wildebeests of course gives them a rather prodigious total range). And none of the alcelaphine fossil species lasts for more than a few million years. Then too, there seem always to have been more than one or two species of the hartebeest-wildebeest gang around: Vrba counts from four to seven species living in the same area at any one time in the fossil record—a situation that has existed for upwards of 5 million years.

A quick check of the behavioral ecology of these beasts confirms what one should by now suspect: impalas are indeed generalists. They occur in a wide variety of habitats, from open plains to wooded grasslands and even forests. They browse on leaves and graze on grass. The alcelaphines graze, and what's more, they are among the antelope world's most finicky eaters, typically preferring a particular species of grass. And alcelaphine species are confined to a much narrower spectrum of habitats than we find impalas ranging through. In a revealing compilation, Vrba obtained accurate wildlife census figures for a number of African parks and game reserves. Impalas tend to outnumber *all* other antelope species combined. The homily is easily inferred: generalists take up a lot more room as individual species than do specialists—and they usually have more organisms per unit area of space than do specialists. But taken all together, the specialist species occupy the same total amount of space. And Vrba concludes that the single impala species is just as successful in contributing genes to succeeding generations as all the diverse alcelaphines (wildebeests and kin) put together. Neither is a "better" strategy: specialists versus generalists—the two strategies amount simply to different modes of environmental perception.

But now we need the macroevolutionary connection—and Vrba has looked at African antelopes from just this perspective. For the specialized Alcelaphini define a trend, rather like the directional increments in brain size in human evolution. In living alcelaphines, there is a spectrum from rather more primitive to rather more derived—from less aberrant-seeming antelopes like bonteboks to the highly derived wildebeests. Sure enough, the bonteboklike alcelaphines appear first in the fossil record, with wildebeests appearing last. There seems to be within alcelaphine history a sort of net movement in the wildebeest direction as time has gone on.

Vrba spotted the trend, and then made a novel suggestion. Favoring the idea that specialists have truly higher rates of speciation, she saw the trend toward increased specialization simply as a result of continued speciation, a willy-nilly accrual of the more and more specialized that simply reflects the intrinsically highly active speciation that goes on within lineages with narrow-niched species. It all boils down to a simple by-product of the physiologies of the organisms themselves: narrowly specialized organisms make up species that tend to fragment easily, setting a train of continued evolutionary activity in motion that continues to produce more and more specialized organisms. In retrospect, we see some apparent directionality and call it a "trend." Perhaps, she thought, there is no need for trends to arise from "species selection."

Vrba noted that Gould and I, S. M. Stanley and others always did tend to emphasize differential species survival in macroevolution. On the basis of her antelope studies, she has contributed two important further ingredients to the "new macroevolution." First, biases in species' births are potentially just as important in producing trends and other phenomena of "macroevolution"—a point raised in a somewhat different context by S. M. Stanley. And, perhaps even more interestingly, she suggested that biases in the births and deaths of entire species may well arise from nothing more exotic than the behaviors, anatomies and physiologies of organisms themselves. And if that is the case, Vrba argued that it is misleading to call the phenomenon "species selection" when the factors really at work influencing the rates of species births and deaths arise *not* from properties of species, but simply from the attributes of organisms. She called her general idea the "effect hypothesis," simply because so much of the comings and goings of entire species, and the biases that create all the apparent order, such as trends, seem to be much more an incidental side effect of organismal biology than any newly discovered force in nature, such as the term "species selection" seems to suggest. I regard her idea as the most interesting and valuable contribution to the paleontological side of modern evolutionary theory to have been proposed—since punctuated equilibria.

But whatever the fate of "species selection" and the "effect hypothesis," we must not lose sight of the real, substantive, empirical gains that we have made—facts about nature that many of us can agree upon. For "species selection" and the "effect hypothesis" are ideas about why we see what we see. Because biologists these days actively debate these notions, leaving an air of uncertainty about the whole matter (scientists are supposed to be dealing with "objective reality" and "truth"—not speculation!), it is well to reiterate one simple generalization: differential species births and deaths are a fact of biological history. Whatever the cause, simply as a descriptive pattern, differential births and deaths of entire species provide a much closer fit to the details of all those large-scale phenomena of evolution—particularly those directional trends, which the synthesis explained simply by invoking the modification of organismic adaptation by natural selection as the ages roll by. Treating species as actors in their own right in the evolutionary drama is now openly accepted as an added element to be reckoned with by all who would seriously consider themselves general theorists of the evolutionary process. Those relatively few vocal dissenters within paleontological ranks have been more than offset by the growing number of geneticists who have come to acknowledge that the comings and goings of entire species are a real phenomenon, one that has had a real role in shaping the history of life—and a phenomenon not really explicitly addressed by traditional evolutionary theory. And this appears to be the main gift of lasting value that punctuated equilibria has contributed to evolutionary biology.

PREDICTIVITY, SCIENCE AND MACROEVOLUTION

It is often said that prediction is the very hallmark of science—indeed, that a science isn't really a science if it lacks the power to predict. The complexities of the evolutionary process being what they are, and as contingent as future biological events are on what will happen in the physical realm, which is equally tough to anticipate, most biologists long ago gave up any hope of foretelling the evolutionary future (though I must say that with stasis so prevalent in the histories of species, it is safe to see little significant anatomical modification in the tea leaves for any particular species alive today). When change *does* occur, precisely what form it will take within certain obvious limitations is anyone's guess.

Thus we tend to make excuses for evolutionary biology—calling it a "historical science" somehow to be excused from having to sport all the trappings of a truly "hard" science. But as we have already seen in another context, that is not what predictability really is all about in science. The point is to test—simply to evaluate—hypotheses someone puts forth to explain some phenomenon or other. And it is merely one way the human mind can grapple with such an order of business when we say: well, if this idea is correct, that means we should expect to observe certain effects in the material world. Thus, if evolution is "true," it implies that all organisms are descended from a single common ancestor, and that we should expect to find a single complex network of similarity interlinking all forms of life. That is what we observe—which doesn't prove evolution, but abundantly corroborates the general notion.

The predictions that have sprung from the adaptive-landscape pictorial metaphor of life's history are really rather vague. Yes, we might include the expectation that organisms will gradually change to fit changing times as a prediction, and assert, as I have asserted, that what we know organisms *really* do when change occurs essentially falsifies the general theory. But just as some paleontologists claim that stasis does not support the notion that speciation is important in the evolutionary game (stasis to them being "zero-rate gradualism"), so too can an ardent supporter of the adaptive imagery of the history of life wriggle out of difficulties by claiming that *both* habitat tracking and change in response to a modified environment occur. And they can always fall back on the "gaps in the fossil record" argument, and deny that the record is in general good enough to tell us anything about the pace of evolutionary change. There really are no hard-and-fast, definite predictions that flow from the generalized adaptation-through-natural-selection-does-everything theory.

But just as I observed when recounting Vrba's antelope analysis, there is an almost rhythmic repetition, a correlation of elements of patterns that we come to expect to recur time after time. Recall that F. J. Teggart, the historian, was exasperated with biologists and social evolutionists alike because of their systematic evasion of what Teggart called the "events" of history. Proffering a view of both human and biological history that he claimed was a far better fit to the facts of the matter—a notion very similar in many details to the outlines of punctuated equilibria—Teggart of course thought he had succeeded in correcting the ills inherent in "evolutionism" that centered on gradual, progressive, adaptive change. Yet how does one go about framing a theory of historical process that really does speak to the historical events in the history of life—or, for that matter, human sociocultural systems?

It is the aim, I think, of all science to make "lawlike" statements, generalizations about the nature of the material universe and why it is the way it appears to be. Most laws are descriptions of what happens given the existence of a set of conditions. Thus, while we might tend to think of gravity as some sort of process (gravity is one of the four basic "forces" physicists investigate), Newton's and all subsequent "laws" of gravity are simply generalizations about just what will happen given the existence of two hunks of matter of specified mass and distance apart. The horror of "historical" science is that no comparable sorts of generalized descriptions seem possible because we are immersed (overwhelmed may seem more like it) in a world of particulars—the events, the minutiae of history. After all, each event in history is unique. Each blade of grass, each individual trilobite, each little local ecosystem. If we try to heed Teggart, we are lost in a welter of particulars, if not wholly overwhelmed by petty detail. This sort of frustration makes the sweeping, eventless generalizations—such as the imagery of the adaptive landscape—less difficult to comprehend. Such models are brave attempts to generalize, to strike at the shackles of particularism to reach the truly general, lawlike statements about evolution.

But Teggart was no devotee of historical particularism. Indeed, Toynbee credits Teggart with the inspiration for comparative history, in which historical generalizations emerge as patterns-in-common. And that really is the point: the way we can pay better attention to the events of biological history, thus make our statements more accurate and perhaps more susceptible to evaluation by consulting Mother Nature, and yet still hope to make truly general statements about the nature of things, is to look for entire classes of events. That's what these *patterns* are: classes, or categories, of events. So far in this chapter, I have touched on a number of cases all manifesting the same concatenation of factors and attributes of organisms and species. Some of these examples involved fish, others snails and mammals. Some dealt primarily with fossils, others mostly with creatures still alive today, while others—especially Vrba's antelopes—included a healthy mixture of the two. Yet there is a certain sameness to all of them, so much so that I dubbed this general class of events "Williams' pattern." It is through such patterns that we can reach some actual generalizations about macroevolution. And it is in the repetition of its components that a pattern can contribute to a macroevolutionary theory by providing us the wherewithal to make predictions and evaluate them.

In 1980, I published a technical book with my colleague Joel Cracraft. In this work, primarily a discourse on how one goes about reconstructing the history of life (and making classifications based on the results), Cracraft and I tried to formulate a general testable macroevolutionary theory based for the most part on the ecological notion of comparative niche width and all the ancillary items of Williams' pattern. We began by specifying four very general sorts of macroevolutionary phenomena. The first was trends. Trends are particularly compelling because of the paradox Gould and I pointed to in 1972. They demand on the face of it an alteration of existing theory.

But there are other phenomena as well—"adaptive radiations," for one. Though there are some native rodents and bats in Australia, marsupials have pretty much had the place to themselves as far as mammalian life is concerned. The top carnivore in the Indonesian archipelago east of Bali (where, as the Balinese still say, "the tiger ends") is a huge monitor lizard ("Komodo dragon") on two of the islands. Elsewhere, where there are no large mammalian carnivores, such as on Sulawesi (Celebes), giant pythons fill the bill. Twenty million years ago an enormous monitor lizard roamed Australia; but for the most part, marsupials have supplied the wolf- and catlike carnivore niche with suitable occupants. And they likewise furnished a vast array of herbivores. Not all the Australian marsupials are close ecological analogues to placental mammals living elsewhere, but the resemblance is striking and drives home the limitations in adaptive design: for mammals, there really are but a few ways to be an effective carnivore, and these general strategies (be they pack hunting, ambushing or solitary stalking by night) bespeak a certain narrow choice of anatomical and behavioral designs.

But it is worth a closer look at the details of these "adaptive radiations." They make much more sense as a rapid and erratic series of bursts, a proliferation from a single common ancestor (which originally occupied, of course, but a single ecological niche). This is what happened to my trilobites in the Devonian of the southern hemisphere. There was an initial burst, then a lot of extinction, followed by a second, much larger burst derived from a small segment of the earlier radiation. There was then a long phase of background ticking over, in which speciation seems to have exactly balanced extinction. Some branches became ever more specialized (to judge from their anatomies), and these are the very branches which produced trends through time, just as Vrba saw happening in one lineage within her antelope group. An "adaptive radiation" is no one evolutionary phenomenon, but a commingling of different rates of speciation and extinction that varies at different times and in different segments of the lineage as time goes on. Within an "adaptive radiation," some sublineages proliferate a lot, while others hardly do much of anything.

As the four basic macroevolutionary patterns, Cracraft and I listed trends, adaptive radiation, arrested evolution and what we called "steady state"—the general pattern of "business as usual," with no net increase or decrease in the average number of species within a lineage, and with no marked accumulation of change in the general appearance of the organisms through time. It now seems that we could have simplified our characterization of macroevolutionary patterns even further. Trends seem, universally, to result from the differential production and survival of the ever-more-specialized within lineages of relatively narrow-niched creatures. "Arrested evolution," on the other hand, in which organisms display little or virtually *no* anatomical change since the inception of their lineage (in some cases, periods spanning several hundred million years), is, as a rule, the flip side of Vrba's antelope example: if species in a lineage remain ecological generalists, the lineage invariably will produce new species at a low rate, and little or no adaptive change will accrue. Today's impalas look and presumably behave very much like those living 5 million years ago. Living fossils—the semilurid name sometimes slapped on these products of "arrested evolution"—are not the same species somehow managing to hang on for truly prodigious periods of time; they are merely the current members of lineages in which both speciation and extinction have been very slow. Such species are typically generalists, and when speciation does occur, little anatomical change seems to go along with it. "Steady state" (in which speciation and extinction rates seem to be balanced but neither abnormally fast nor abnormally slow) and "adaptive radiations" seem, in contrast to the other two patterns, rather general sorts of phenomena: I have already remarked that the radiation of the southern Devonian calmoniid trilobites, when examined in detail, breaks down into some trends, some low-rate lines and some "status quo" patterns. And all of this looks like patterns of differential speciation and species extinction—biases in the births and deaths of species within lineages of differing ecological predilections.

But how do we test such notions? It is best, of course, to have members of a lineage still alive so we can understand in depth the nature of niche exploitation, as Vrba was able to do with her antelopes. But even with fossils there are a few things we can always do. The situation is best when we have a pair of lineages arising from a common ancestor and proceeding at different rates—again, like the impalas and wildebeests in Africa. If we spot a trend, say, in the fossil record, we would expect that the lineage would typically sport a number of species living in the same general region at any one time. We would expect the lineage to display an accumulation of specialization through time. Each species should be relatively short-lived and restricted to some relatively narrow area of the total geographic range of the group. Habitat preferences should be fairly narrow—as judged, for example, from the range of environments detectable from the sediments enclosing the fossils and the spectrum of other sorts of organisms found along with the fossils. For cases of living fossils, we would expect them to be generalists, anatomically unspecialized and ranging into a broader array of environments, spread out over wide areas and persisting through longer chunks of geological time.

It seems an open-and-shut case: all the examples I have discussed in this chapter fill the bill nicely, so it seems that not only am I proclaiming testability, but the nice, tight little argument we can weave around "Williams' pattern" as the key to understanding macroevolutionary patterns emerges as simultaneously tested and triumphantly confirmed. So we'll close this discussion of macroevolution with a case that *doesn't* fit the model, a case that nicely illustrates the difficulties still inherent in the attempt to frame generalizations in macroevolutionary theory. Consider human evolution—specifically that trend toward increase in brain size within our lineage. If ever there has been a broad-niched species, it is *Homo sapiens*. And many of the earmarks of generalists show up in the human story. If we follow the outline of human evolution that Ian Tattersall and I sketched in our *Myths of Human Evolution*, we can admit *Homo erectus*, *Homo habilis* and finally *Australopithecus africanus* to our own subset of the hominid family, a scheme that excludes only the robust neanderthals and robust australopithecines from consideration as species ancestral to our own at some point in time. *Australopithecus africanus* and *Homo erectus* each remained stable for more than a million years; considering the rapid overall pace of human evolution, such stability agrees pretty well with our expectation of the evolutionary "behavior" of ecological generalists. *We*, as a species, might hope to do as well. And within this segment of the hominid tree, only one species seems to have been present at any one time in the Old World, certainly as an overwhelming generalization: another corroboration of the notion that ours is a broad-niched breed. And in general, we are anatomically rather primitive primates. Even our full bipedalism, which is certainly a specialization, is by now an old invention: our locomotory anatomy hasn't changed a bit for 3.5 million years—even more support for our generalist characterization.

Yet the pace of anatomical change in the brain has been prodigious. And the trail that change has left in the rocks is one of episodic but unidirectional transformation: a *bona fide* trend. One doesn't expect generalists to show a pattern of rapid directional evolutionary change!

Well, we have always thought of ourselves as rather a peculiar species. Though there may well be other egregious exceptions to Williams' pattern out there, it is somewhat comforting that it is our own lineage which throws the spanner into the macroevolutionary works. But perhaps human evolution is the exception that in the end proves the rule: for our sociocultural systems really are a highly complex evolutionary specialization. And these are rooted, in some arguable fashion, in our anatomical brains. We are, in short, more of a hybrid of generalist/specialist than most other creatures appear to be. And if this is true, perhaps we can better understand the curious mixture of pattern that our own evolutionary history seems to hold.

So there it is: punctuated equilibria–inspired macroevolutionary theory, warts and all. It isn't particularly neat, elegant, all-embracing, completely testable—or even as yet totally thought through. It *does* give a better fit to the kinds of events that typify real evolutionary case histories, and on those grounds alone represents a distinct improvement over the simpler and probably more elegant formulations of the modern synthesis. What we have gotten from punctuated equilibria, more than anything else, is the renewed urge to think about large-scale evolutionary patterns, and to pursue this thought along lines that remained unexplored under the older paradigm that focused just on adaptive change. If in the end the next generation will have to discard it all (or only part of it, as I, being human, would wish) to clear the way for an even better description of nature, so be it. That's what science—any science—is all about.

7 WHERE TO? WHAT NEXT?

IN ESMERALDA COUNTY, Nevada, just east of the California border, what's left of the advent of complex life is there in the rocks for all who are curious. Now a barren and rather desolate place, with dry scrubland lying among rocky hills, it presents a seeming irony as a locale where the beginnings of life's big explosion would crop up. And yet the Joshua trees, hawks and prairie rattlers, the coyotes and mule deer are all here; the place is by no means lifeless. And just a few miles away, in the White-Inyo Mountains, stand the bristlecone pines, those hardy, gnarled trees said to be the longest-lived of all organisms on earth.

History, Teggart told us back in 1925, is the explanation of "how things have come to be as we see them today." In the outcrops of Esmeralda County, large cabbage-shaped mounds of finely laminated limy sediment, trapped by filaments of a blue-green alga (which is a bacterium, really) are the most conspicuous evidence of simple life forms. The creatures momentarily perched on top of those outcrops—birds, mammals, lizards, insects—are radically different from those simple bacteria. They are all eucaryotes, with much more complex cellular structure. And their bodies consist of up to 100 or so different sorts of cells.

Someone equipped with a powerful imagination could perhaps fill in the gaps between these 600-million-year-old cabbagy stromatolites and the modern forms of life that succeed them in Nevada. Fortunately we need not be content with just this approach to Teggart's problem: Nevada, after all, isn't the whole world. Most of the larger groups of organisms, the bigger divisions of life, that have ever appeared are still with us today. Stromatolites, which go back in the record more than 3 billion years, are alive and well in Shark Bay in northern Australia. The spectrum of organisms we see around us today fills in that gap between past and present in Nevada in a very general sort of way. The old notion of the Great Chain of Being reflected some basic truths in a time when the idea of evolution had yet to gain legitimacy. But this chain, this *Scala Naturae*, goes back as an idea—an observation, really—to Aristotle. There really is an interconnectedness to all life.

But the modern world tells us only the general outline of how all present-day creatures are interrelated. It doesn't really tell us much in detail about what the historical events were that shaped the world as we find it today. That's where the fossil record and earth history step in: as fragmented as the story is, as incomplete as the documentation must needs be, it is a temporal ledger of organic history.

Paleontologists have thought this way for years. We have long supported the myth that only paleontologists are actually privy to the history of life: climb enough hills, crack enough rocks and all will be revealed. But the analysis of just how all creatures really are interrelated relies heavily on detailed study of organisms still living—simply because there is so much more about the basic biology of living creatures that we can learn than we can possibly say about fossils. And much the same is true for gaining ground in the long struggle to learn *how* life has evolved. For one thing is clear: no matter how we derive a picture of *what* has happened in the history of life, that picture alone will yield no sure, complete theory about the nature of the evolutionary process.

Yet Teggart was dead right about the perils of "evolutionism." The theory that evolution takes place as adaptive modifications in small steps, if not always even in pace, took absolute precedence over empirical experience with the actual events in the history of life. If it is dangerous to extrapolate from a 600-million-year-old bacterium to the hawk roosting on it today, it must also be risky to assert that the genetic processes we can observe going on within human lifetimes—processes that can and do modify the gene frequencies, hence observable characteristics, of organisms within populations—are all we need consider to explain that history of life. But the problem really is tricky: for there is no doubt, at least in my own mind, that organisms in all the species that have ever lived are adapted to

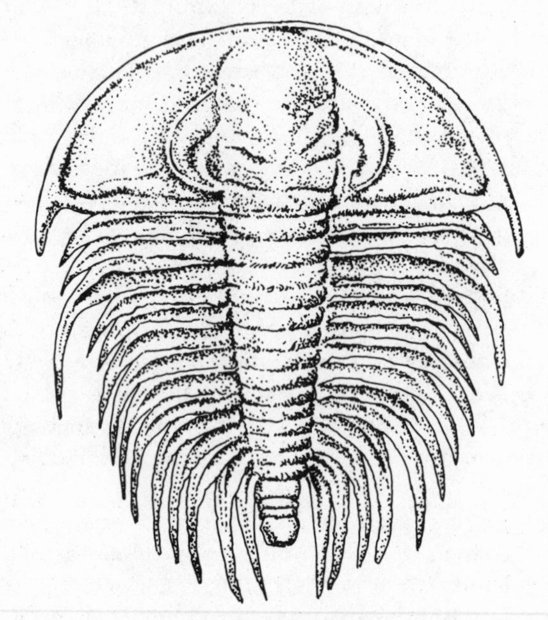

Olenellid trilobite.

their environments. Thus the features, the anatomical characteristics, of these organisms that we see occasionally changing through time represent, at least for the most part, modified adaptations. We go to the genetics lab and observe that we have a perfectly good theory—selection modifying gene composition in response to a stimulus—and we assert, logically enough, that we need look no further for a basic understanding of the principles that have governed the history of life.

And there is really nothing wrong with this line of thought—as far as it goes. Not too far above the zone of stromatolites, a little higher in the general rock sequence at the Nevada–California border, olenellid trilobites appear. Until recently, they were the oldest truly complex forms of life known. (They still rank among the oldest hard-shelled invertebrates to show up in the fossil record.) Olenellids are appropriately primitive: as the first of the trilobites, the beginning of a history that lasted a third of a billion years, they really *ought* to be primitive. Their tails are rather wormlike: the body segments just keep getting narrower and shorter all the way back, with a little terminal nub bringing up the rear. They have no great, fused tailpiece which virtually all other trilobites have, a body unit that really stamps all later trilobites as unique within the arthropod realm. But the *rest* of the olenellid body is something else again: with the head supporting large, crescentic eyes, widening into a central region that evidently housed a rather large stomach and having spines projecting from its side, the olenellids resemble in a superficial, yet compelling, way my Gondwana trilobites which lived nearly 200 million years later. Olenellids too were fairly large—anywhere from an inch or two up to half a foot or more in length. Most trilobites that succeeded them were smaller.

Now, I have already admitted my ignorance concerning the adaptive significance of most features of trilobite anatomy. I really don't know why they reached the sizes they did, why those spines stuck out of the sides of their head or what kind of food they stuffed into their enlarged stomachs. It would be nice to know. But it is rather interesting that trilobites time and again struck upon the same basic body plans, proffering the same basic *Gestalt*, in groups only remotely related to one another. In other words, it seems a near certainty that the Cambrian olenellids had a general mode of life that was exploited once again, hundreds of millions of years later, by the Gondwana trilobites—whose most immediate ancestors did not particularly resemble olenellids at all. Repetition of adaptive themes seems very definitely to be a strong signal in the history of life.

And if that is so, then the case for adaptation through natural selection as the general theory of the history of life seems, of course, strengthened. It is commonly assumed that punctuated equilibria somehow undermines the notion of adaptation. But our message is not that adaptation somehow doesn't exist; that would be silly. Our message on adaptation is really twofold.

The geometry of the history of life, the actual pulse of change, does not support the conventional supposition, which saw a simple extrapolation from the genetics lab to the fossil fields of Nevada. Adaptation remains—but stability is a stronger signal than change, and that stability makes sense only in a context of adaptation. If you really have built a better mousetrap, for heaven's sake don't fool around with it.

Our second message, in its most general manifestation, is a bit more subtle. It is really a statement about complexity. We are saying, in effect, that the world, the living world today *and* its past history, is more complicated than existing neo-Darwinian theory will acknowledge. If punctuated equilibria were only the proposition that rates of anatomical change vary widely, and systematically, such that change occurred only in relatively rapid bursts, there would be no real need to modify existing theory. To this very day, the major criticism leveled against us is precisely that: we have set up a straw man in claiming that Darwin (or any of his main descendants) insisted that evolution was always slow, steady and gradual. There is room for variation in rates, Darwin knew it and, what's more, the synthetic theory, that blend of genetics and Darwinian outlook, is equipped to embrace such a contingency. In that case, all that punctuated equilibria contributes is a renewed appreciation for stasis—long-term stability—known to Darwin's contemporaries (paleontologists) and even to Darwin himself, but still denied (ironically) by some of Darwin's supposed "defenders" today. If this is all there is to it, the modern synthesis is hardly shaken.

Yet in linking these patterns of change with speciation—the origin of new reproductive communities from old—in insisting, in short, that most adaptive change appears to go hand in hand with the origin of reproductively isolated groups, we drew attention to a duality that has permeated everyone's evolutionary thinking since Darwin, a duality glossed over to such an extent in evolutionary theory that it threatens to blow the lid off the modern synthesis.

For on the one hand we have reproductive success and on the other we have economic adaptation. Natural selection is the idea that economically more successful organisms will survive and reproduce and, on average, leave more offspring to populate the succeeding generation than will their less economically viable competitors. With organisms, the idea seems to work, as obviously troublesome as some of our own common human experiences with the world show it to be. Serendipity, for instance, enters in quite a lot when it comes to reproduction. And newspapers have carried the latest instance of a long-standing concern: the Mayor of Singapore recently became alarmed over some statistics which seemed to show that the economically more viable citizens of that city-state were having fewer children, on average, than the less well-to-do. Echoes of an era! Human social organization, with its complex commingling of sex and economics, is not a sure guide to the general situation among sexual organisms in nature. But social Darwinists have been struck for years by the seeming incongruity in which "economic success" is often inversely proportional to "reproductive success." (A partial answer to the riddle, at least in technologically advanced societies, is not that natural selection is thereby refuted—merely that economically disadvantaged classes realize a form of wealth in children, a wealth not measured solely in emotional terms.)

Natural selection works because organisms have dual identities: they are economic machines, and they reproduce. The connection, in most organisms, between the two functions is remote: thus selection works only as a statistically averaged phenomenon within populations. Now consider the support that stasis—species stability—lends to the argument that species are *individuals*, historical entities that originate, live on awhile and ultimately die. From time to time, as we have seen, these entities, with an internal fabric as fraught with space as a chunk of Swiss cheese, give rise to other, new reproductive communities.

The possibility that there exist these other, larger-scale entities is perhaps the major gift of punctuated equilibria to the ongoing task of learning more about how the evolutionary process really works. The idea itself—that there are large-scale entities so vast in space and time that we, locked into our humanly scaled perceptions, are scarcely aware of them—has been around for a while. But it forms no functional part of conventional evolutionary theory, which deals only with genes, organisms, populations—and to some narrow extent, species. Punctuated equilibria above all else suggests that species really do have a concrete reality. Species really are actors in the evolutionary drama.

Well, if this be arguably so for species, why not for other, similarly large and diffuse biological "entities" as well? Why not, indeed? The prospect is very exciting, and though still very much embryonic, has already led to some initial theoretical exploration. For example, species are conventionally acknowledged to cluster into larger groups—genera, families and so forth. The arrangement, the Linnaean hierarchy, mirrors that nested pattern of evolutionary novelties, those adaptations, as Darwin pointed out so long ago. Long alleged to be simply artifacts of our unquenchable thirst to classify, to impose order on nature, these clusters of related species are now once again coming to be seen literally as branches of various sizes on the very real "tree of life." Mammals are a real group: we are fairly sure this is so because all mammals have hair, three middle-ear bones and other telltale signs of descent from some remote ancestral species. It is simply now coming back into vogue, after a long hiatus, to see such entities as mammals as having themselves, as a unit, an actual beginning, a real history—and, inevitably, an end (trilobites, after all, have been gone a long time). These larger groups, then, are also players in the evolutionary game.

But there are other large-scale biological entities as well, other candidates for the mantle of "evolutionary unit." And these lie in the ecological sphere. Ecologists call local aggregates of organisms all of the same species "populations." Mixtures of populations of different organisms form "communities"—and when such clusters of different populations are considered along with the physical environment in which they live, and when we think of the dynamics of energy flow within that web, we speak of "local ecosystems." Ecologists differ among themselves on this point, but a number of them insist that communities and ecosystems don't really exist in any real sense. They are, instead, ephemeral collections of populations, in constant flux and with no real definable boundaries.

Where have we heard *that* before? It is precisely the same description that biologists have been overly fond of when confronting species. And once again the fossil record seems to support ecologists who see it otherwise. Fossil communities are as incomplete as is the record through time for any one particular species. For one thing, soft-bodied organisms as a rule aren't preserved at all. But when we compare, say, a collection of hard-shelled invertebrates from the Pliocene Purisima Formation at Capitola with the modern hard-shelled invertebrate community still living in Monterey Bay, we see (as I have been emphasizing) great species stability. But we also see a tremendously conservative ecological organization: the very fabric of economic life in shallow-water central-Californian marine ecosystems seems to have been altered only very slightly over the last 3 or 4 million years. This is what Coope saw in his Pleistocene beetles as well. And when we look at Paleozoic communities, again a great sameness prevails: even if the *species* change—say, when we compare brachiopod communities 100 million years apart—we see there is still more than a mere semblance of similarity: it is as if the characters, the species, change their identity, but the same genera, or families, are still there. There is a hierarchy of ecological similarity very much like the hierarchy of adaptive, structural similarity that unites all of life.

Now, if such items—larger taxonomic groups and larger ecological units—really are to be construed as historical entities, what about that dualism between economics and reproduction, a dualism that makes natural selection an inevitability as a force in nature, and which, at the same time, ironically makes it a less than rigidly deterministic factor in evolutionary stasis and change? Here, I think, we come to an interesting frontier: some of these larger units, such as communities, are patently economic entities. Others seem more reproductive in nature. Remember our very definition of species: they are coherent reproductive communities. The interplay between the economic realm and the reproductive—when those functions are segregated into different entities, not embodied in the same unit as they are (uniquely) in organisms—makes for a fascinating series of possibilities, possibilities only now being addressed in the technical literature.

But the realm of human thinking about evolution is widening. And there is the embodiment of an enhanced appreciation of complexity: more than genes and organisms are at play in the evolutionary arena. And finally, we can get a glimpse of just how it is that ecology—economics—is involved in the evolutionary process.

What this means on a more pragmatic level is the fulfillment of yet another promise of punctuated equilibria. One simply cannot go to Esmeralda County and think rationally of those stromatolites and the advent of olenellid trilobites in terms of shifting gene frequencies. Yes, those organisms had genes. Yes, genetic change underlay the transformations in organic form that occurred through time—gross versions of which show up in those outcrops. But the distance between genes and 570-million-year old trilobites is too great for anyone to be able to match them up.

Yet we really must somehow match up our grasp of life's history with our theory of how life has evolved. Teggart was right: the events of life's history somehow must hold our theory of evolution accountable. The old paleontological reaction to such frustration was to throw out genetics—or invent a seemingly more suitable genetic theory (as Osborn did). But that is even less rational than trying to see genes in the fossil record.

What to do? The story of punctuated equilibria is a small example of what really can be done. We paid attention to actual patterns of stasis and change that really do seem to be general in the fossil record. We happened to look at small-scale changes within and between species, and to interpret them in a way that fitted speciation theory better than the alternative model of adaptation-through-natural-selection. But that is only one example. If there truly are these large, hard-to-perceive entities—these species, larger taxa, communities and ecosystems—it turns out that the fossil record, by its very coarseness of scale and, of course, its element of time, gives us entry into a world if only we look at it right—a world otherwise invisible, as the history of biology so eloquently attests.

The Devonian rocks of the central Andes are 10,000 feet thick—two miles of sediment that record perhaps 25 million years of time. That is pretty coarse: 1 foot of rock averages out to 2,500 years. If evolution is only generation-by-generation change in gene frequency, then despair is appropriate if we wish to translate theory into an understanding of organic change up through these rocks. And 10,000 feet is a whacking great chunk; most sediments span comparable time intervals with far less muds, silts and sands accumulated to encase a fossil record. But what if we truly are dealing with entities that spread out over major masses of continents and oceans, entities that go on living for millions of years? What if—as we already know must be true—the ecological history of life reflects a shuffling of species? What then? How can we come to grips with such enormous and diffuse objects? Where is our reverse telescope? The answer is clear: if there is any reason to expect some such entities to behave as though they were coherent historical units, our initial despair over the fossil record gives way to elation: it is wonderful to be able to climb a 500-foot-high hillside and pass through 10 million years of time. For only in such circumstances can any single human being harbor any hope at all of comprehending the wholeness of these vast organic entities. It is only then that we can begin to grasp the truly coarse, large-scale processes and events that go on in organic nature, processes every bit as important as the shifting gene frequencies in populations—or, on an even smaller scale, the jumping of genes on the molecular level within individual cells.

At the core of punctuated equilibria lies an empirical observation: once evolved, species tend to remain remarkably stable, recognizable entities for millions of years. The observation is by no means new: nearly every paleontologist who reviewed Darwin's *Origin of Species* pointed to his evasion of this salient feature of the fossil record. But stasis was conveniently dropped as a feature of life's history to be reckoned with in evolutionary biology. And stasis had continued to be ignored until Gould and I showed that such stability is a real aspect of life's history which must be confronted—and that, in fact, it posed no fundamental threat to the basic notion of evolution itself. For that was Darwin's problem: to establish the plausibility of the very idea of evolution, Darwin felt that he had to undermine the older (and ultimately biblically based) doctrine of species fixity. Stasis, to Darwin, was an ugly inconvenience.

Yet it need not be. Observations of radically different rates of evolutionary change can be explained in any number of ways. Darwin himself was aware of fluctuations of rates of change within single lineages, and modern theories of population genetics are well equipped to explain how directional natural selection and other parameters may be expected to vary, and how their variation might result in a spectrum of rates of change.

Yet Gould and I insisted at the outset that another aspect of the evolutionary process was at work producing those quick (geologically speaking) bursts of change, followed by those protracted periods of business-as-usual nonchange. We thought that such patterns, particularly when drawn on a map of distribution of a group of fossils, agreed pretty well with the standard notions of geographic speciation. The essential idea here is that new species—new reproductive communities—tend to bud off in some isolated region from a more widely spread ancestral species. Thus a second element is added to the simple notion of adaptive change through natural selection: the concept of the fragmentation of reproductive communities. And once it is realized that adaptations are likely to remain stable as long as a species can continue to exploit the familiar conditions of its habitat, stasis no longer seems so very difficult to understand.

But we also saw that *if* those bursts of change really were produced by the budding off of descendant from parental species, there was a corollary: adaptive change in evolution is not a pervasive, ongoing process, but rather seems concentrated at those speciation events. The time scale involved—5,000 to 50,000 years, on average—seems about right given what we know about both speciation and change possible under natural selection in modern organisms. Thus we were able to effect a reorientation in the minds of most of our colleagues: Change, especially of the gradual, progressive sort, is by no means an inevitability given the mere passage of time. Empirically, change comes in bursts, and these bursts probably represent speciation "events," the births of new species. More difficult for our colleagues to accept is the corollary—that adaptive change (however great or trivial) seems to occur in conjunction with speciation. Indeed, I would say that change seems at least as much a by-product as the cause of the birth of a new species.

Thus speciation is now better melded into mainstream paleontological thinking on evolution. But so far, we are still faced with a one-way street. We paleontologists make our observations and utilize biological theory to explain them. Thus in rejecting a notion of pervasive natural selection gradually modifying an entire species to yield gradual, progressive change, we simply substitute an alternative item from the catalogue of modern evolutionary biology: speciation theory. We have perhaps effected a better fit between theory and history; but we really haven't as yet contributed to an improved understanding of the mechanics of the evolutionary process.

So much for the real, but finite, value of punctuated equilibria. But its contribution to a deeper understanding of the workings of evolution is just beginning. For Gould and I took seriously the notion that species are real entities, a notion that I have explored in some depth in Chapter 4. Though there is some precedent for looking at species this way, it remains for most modern biologists a truly "radical" proposition (as biologist Michael Ghiselin terms it). It turns out, moreover, that the fossil record is the best place to see the *larger* biological entities—not just species, but entire natural groups of creatures, like "mammals," and also large-scale ecological units. If we admit that such entities have had histories, plus mechanisms of birth and death, stasis and change, then evolution simply cannot be entirely a matter of shifting gene representations from one generation to the next. Gould and I discussed the possibility of nonrandom species representation through time within a lineage—the notion now known as "species selection." An entirely new field of active inquiry into "macroevolution" has sprung up—if not entirely to be traced to the original exposition of punctuated equilibria, at least partly beholden to it for its genesis.

And precisely the same sort of activity—based on the recognition that there are biological *entities* out there which manifestly are involved with the evolutionary process, yet not formally a part of evolutionary theory—is going on elsewhere in biology. The best example is the revelation of the complex organization of genetic elements even in the simplest of organisms. The diversity of these sorts of genetic constituents, whereby some genetic molecular bits can regulate the activities of others, some can apparently "jump" around, and some bias their own representation in succeeding generations, shows that there is a lower, molecular level to the evolutionary process which is not at all fully understood, nor is it directly addressed by existing theory.

It is this promise—that these lower-level molecular units, along with the larger-scale units best seen in the fossil record, actually do exist and play active roles in the evolutionary process—which reshapes the role paleontology and other disciplines will play in evolutionary biology in the years ahead. This is, to be sure, controversial stuff. But it is being taken seriously, and *that* is punctuated equilibria's greatest reward.

APPENDIX: PUNCTUATED EQUILIBRIA: AN ALTERNATIVE TO PHYLETIC GRADUALISM

Niles Eldredge and Stephen Jay Gould

(First published in *Models in Paleobiology*,
T. J. M. Schopf, editor, 1972)

STATEMENT

In this paper we shall argue:

(1) The expectations of theory color perception to such a degree that new notions seldom arise from facts collected under the influence of old pictures of the world. New pictures must cast their influence before facts can be seen in different perspective.

(2) Paleontology's view of speciation has been dominated by the picture of "phyletic gradualism." It holds that new species arise from the slow and steady transformation of entire populations. Under its influence, we seek unbroken fossil series linking two forms by insensible gradation as the only complete mirror of Darwinian processes; we ascribe all breaks to imperfections in the record.

(3) The theory of allopatric (or geographic) speciation suggests a different interpretation of paleontological data. If new species arise very rapidly in small, peripherally isolated local populations, then the great expectation of insensibly graded fossil sequences is a chimera. A new species does not evolve in the area of its ancestors; it does not arise from the slow transformation of all its forebears. Many breaks in the fossil record are real.

(4) The history of life is more adequately represented by a picture of "punctuated equilibria" than by the notion of phyletic gradualism. The history of evolution is not one of stately unfolding, but a story of homeostatic equilibria, disturbed only "rarely" (i.e., rather often in the fullness of time) by rapid and episodic events of speciation.

THE CLOVEN HOOFPRINT OF THEORY

Innocent, unbiased observation is a myth.
—P. B. MEDAWAR, 1969, p. 28

Isaac Newton possessed no special flair for the turning of phrases. Yet two of his epigrams have been widely cited as guides for the humble and proper scientist—his remark in a letter of 1675 written to Hooke: "If I have seen farther, it is by standing on the shoulders of giants," and his confusing comment of the *Principia* (1726 edition, p. 530): "hypotheses non fingo"—[I frame no hypotheses]. The first is not his own; it has a pedigree extending back at least to Bernard of Chartres in 1126 (Merton, 1965). The second is his indeed, but modern philosophers have offered as many interpretations for it as the higher critics heaped upon Genesis 1 in their heyday (see Mandelbaum, 1964, p. 72 for a bibliography).

Although most scholars would now hold, with Hanson (1969, 1970, see also Koyré, 1968), that Newton meant only to eschew idle speculation and untestable opinion, his phrase has traditionally been interpreted in another light—as the credo of an inductivist philosophy that views "objective" fact as the primary input to science and theory as the generalization of this unsullied information. For example, Ernst Mach, the great physicist-philosopher, wrote (1893, p. 193): "Newton's reiterated and emphatic protestations that he is not concerned with hypotheses as to the causes of phenomena, but has simply to do with the investigation and transformed statement of *actual facts* . . . stamps him as a philosopher of the *highest* rank."

Today, most philosophers and psychologists would brand the inductivist credo as naive and untenable on two counts:

(1) We do not encounter facts as *data* (literally "given") discovered objectively. All observation is colored by theory and expectation. (See Vernon, 1966, on the relation between expectation and perception. For a radical view, read Feyerabend's (1970) claim that theories act as "party lines" to force observation in preset channels, unrecognized by adherents who think they perceive an objective truth.)

(2) Theory does not develop as a simple and logical extension of observation; it does not arise merely from the patient accumulation of facts. Rather, we observe in order to test hypotheses and examine their consequences. Thus, Hanson (1970, pp. 22–23) writes: "Much recent philosophy of science has been dedicated to disclosing that a 'given' or a 'pure' observation language is a myth-eaten fabric of philosophical fiction. . . . In any observation statement the cloven hoofprint of theory can readily be detected."

Yet, inductivist notions continue to control the methodology and ethic of practicing scientists raised in the tradition of British empiricism. In unguarded moments, great naturalists have correctly attributed their success to skill in hypothesizing and power in imagination; yet, in the delusion of conscious reflection, they have usually ascribed their accomplishments to patient induction. Thus, Darwin, in a statement that should be a motto for all of us (letter to Fawcett, September 18, 1861, quoted in Medawar, 1969), wrote:

> About thirty years ago there was much talk that geologists ought only to observe and not theorize; and I well remember someone saying that at this rate a man might as well go into a gravel-pit and

> count the pebbles and describe the colours. How odd it is that anyone should not see that all observation must be for or against some view if it is to be of any service.

Yet, in traditional obeisance to inductivist tenets, he wrote in his autobiography that he had "worked on true Baconian principles, and without any theory collected facts on a wholesale scale" (see discussion of this point in Ghiselin, 1969a; Medawar, 1969; and de Beer, 1970).

Almost all of us adhere, consciously or unconsciously, to the inductivist methodology. We do not recognize that all our perceptions and descriptions are made in the light of theory. Leopold (1969, p. 12), for example, claimed that he could describe and analyze the aesthetics of rivers "without introduction of any personal preference or bias." He began by generating "uniqueness" values, but abandoned that approach when the sluggish, polluted, murky Little Salmon River scored highest among his samples. He then selected a very small subset of his measures for a simplified type of multivariate scaling. As he must have known before he started, Hells Canyon of the Snake River now ranked best. It cannot be accidental that the article was written by an opponent to applications then before the Federal Power Commission for the damming of Hells Canyon. (It is no less fortuitous that so many philosophers, Hegel and Spencer in particular, generated ideal states by pure reason that mirrored their own so well.)

In paleontology, even the most "objective" undertaking, the "pure" description of fossils, is all the more affected by theory because that theory is unacknowledged. We describe part by part and are led, subtly but surely, to the view that complexity is irreducible. Such description stands against a developing science of form (Gould, 1970a, 1971a) because it both gathers different facts (static states rather than dynamic correlations) and presents contrary comparisons (compendia of differences rather than reductions of complexity to fewer generating factors). D'Arcy Thompson, with his usual insight, wrote of the "pure" taxonomist (1942, p. 1036), "when comparing one organism with another, he describes the differences between them point by point and 'character' by 'character.' If he is from time to time constrained to admit the existence of 'correlation' between characters . . . yet all the while he recognizes this fact of correlation somewhat vaguely, as a phenomenon due to causes which, except in rare instances, he can hardly hope to trace; and he falls readily into the habit of thinking and talking of evolution as though it had proceeded on the lines of his own description, point by point and character by character."

The inductivist view forces us into a vicious circle. A theory often compels us to see the world in its light and support. Yet, we think we see objectively and therefore interpret each new datum as an independent confirmation of our theory. Although our theory may be wrong, we cannot confute it. To extract ourselves from this dilemma, we must bring in a more adequate theory; it will not arise from facts collected in the old way. Paleontology supported creationism in continuing comfort, yet the imposition of Darwinism forced a new, and surely more adequate, interpretation upon old facts. Science progresses more by the introduction of new world-views or "pictures"* than by the steady accumulation of information.

This issue is central to the study of speciation in paleontology. We believe that an inadequate picture has been guiding our thoughts on speciation for 100 years. We hold that its influence has been all the more tenacious because paleontologists, in claiming that they see objectively, have not recognized its guiding sway. We contend that a notion developed elsewhere, the theory of allopatric speciation, supplies a more satisfactory picture for the ordering of paleontological data.

PHYLETIC GRADUALISM: OUR OLD AND PRESENT PICTURE

Je mehr sich das palaeontologische Material vergrössert, desto zahlreicher und vollständiger werden die Formenreihen.

—ZITTEL, 1895, p. 11

Charles Darwin viewed the fossil record more as an embarrassment than as an aid to his theory. Why, he asked (1859, p. 310), do we not find the "infinitely numerous transitional links" that would illustrate the slow and steady operation of natural selection? "Why then is not every geological formation and every stratum full of such intermediate links? Geology assuredly does not reveal any such finely graduated organic chain; and this, perhaps, is the gravest objection which can be urged against my theory" (1859, p. 280). Darwin resolved this dilemma by invoking the great inadequacy of surviving evidence (1859, p. 342): "The geological record is extremely imperfect and this fact will to a large extent explain why we do not find interminable varieties, connecting together all the extinct and existing forms of life by the finest graduated steps. He who rejects these views on the nature of the geological record, will rightly reject my whole theory."

Thus, Darwin set a task for the new science of evolutionary paleontology: to demonstrate evolution, search the fossil record and extract the rare exemplars of Darwinian processes—insensibly graded fossil series, spared somehow from the ravages of decomposition, non-deposition, metamorphism, and tectonism. Neither the simple testimony of change nor the more hopeful discovery of "progress" would do, for anti-evolutionists of the catastrophist schools had claimed these phenomena as consequences of their own theories. The rebuttal of these doctrines and the test for (Darwinian) evolution could only be an *insensibly graded fossil sequence*—this discovery of all transitional forms linking an ancestor with its presumed descen-

* We have no desire to enter the tedious debate over what is, or is not, a "model," "theory," or "paradigm" (Kuhnian, not Rudwickian). In using the neutral word "picture," we trust that readers will understand our concern with alternate ways of seeing the world that render the same facts in *different* ways.

dant (FIGURE *1*). The task that Darwin set has guided our studies of evolution to this day.*

In titling his book *On the Origin of Species by Means of Natural Selection,* Darwin both identified this event as the keystone of evolution and stated his belief in its manner of occurrence. New species can arise in only two ways: by the transformation of an entire population from one state to another (phyletic evolution) or by the splitting of a lineage (speciation). The second process must occur: otherwise there could be no increase in numbers of taxa and life would cease as lineages became extinct. Yet, as Mayr (1959) noted, Darwin muddled this distinction and cast most of his discussion in terms of phyletic evolution. His insistence on insensibly graded sequences among fossils reflects this emphasis, for if species arise by the gradual transformation of entire populations, an even sequence of intermediates should indeed be found. When Darwin did discuss speciation (the splitting of lineages), he continued to look through the glasses of transformation: he saw splitting largely as a sympatric process, proceeding slowly and gradually, and producing progressive divergence between forms. To Darwin, therefore, speciation entailed the same expectation as phyletic evolution: a long and insensibly graded chain of intermediate forms. Our present texts have not abandoned this view (FIGURE 2), although modern biology has.

In this Darwinian perspective, paleontology formulated its picture for the origin of new taxa. This picture, though rarely articulated, is familiar to all of us. We refer to it here as "phyletic gradualism" and identify the following as its tenets:

(1) New species arise by the transformation of an ancestral population into its modified descendants.

(2) The transformation is even and slow.

(3) The transformation involves large numbers, usually the entire ancestral population.

(4) The transformation occurs over all or a large part of the ancestral species' geographic range.

These statements imply several consequences, two of which seem especially important to paleontologists:

(1) Ideally, the fossil record for the origin of a new species should consist of a long sequence of continuous, insensibly graded intermediate forms linking ancestor and descendant.

(2) Morphological breaks in a postulated phyletic sequence are due to imperfections in the geological record.

Under the influence of phyletic gradualism, the rarity of transitional series remains as our persistent bugbear. From the reputable claims of a Cuvier or an Agassiz to the jibes of modern cranks and fundamentalists, it

---

* Beliefs in "saltative" evolution, buttressed by de Vries' "mutation theory," collapsed when population geneticists of the 1930's welded modern genetics and Darwinism into our "synthetic theory" of evolution. The synthetic theory is completely Darwinian in its identification of natural selection as the efficient cause of evolution.

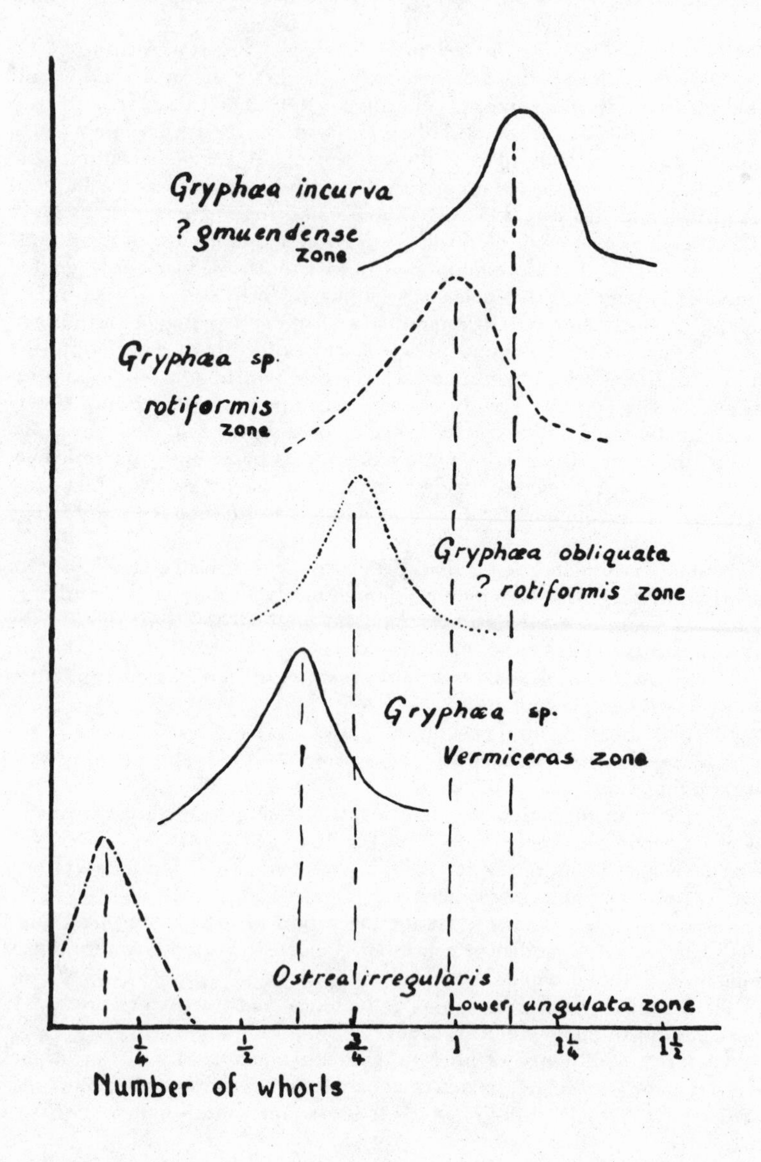

FIGURE *1*: The classic case of postulated phyletic gradualism in paleontology. Slow, progressive, and gradual increase in whorl number in the basal Liassic oyster *Gryphaea*. From Trueman, 1922; figure 5.

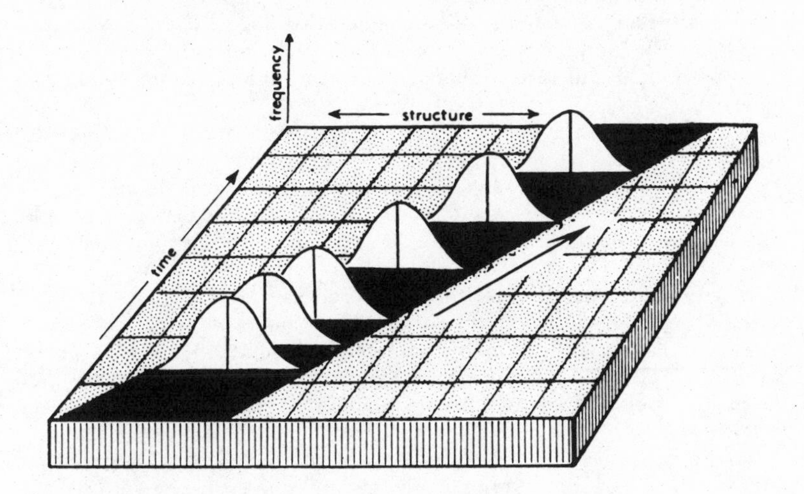

FIGURE 2: A standard textbook view of evolution *via* phyletic gradualism. From Moore, Lalicker, and Fischer, 1952; figure 1–14.

has stood as the bulwark of anti-evolutionist arguments: "For evolution to be true, there had to be thousands, millions of transitional forms making an unbroken chain" (Anon., 1967—from a Jehovah's Witnesses pamphlet).

We have all heard the traditional response so often that it has become imprinted as a catechism that brooks no analysis: the fossil record is extremely imperfect. To cite but one example: "The connection of arbitrarily selected 'species' in a time sequence, in fact their complete continuity with one another, is to be expected in all evolutionary lineages. But, *fortunately*, because of the imperfect preservation of fossil faunas and floras, we shall meet relatively few examples of this, no matter how long paleontology continues" (Eaton, 1970, p. 23—our italics; we are amused by the absurdity of a claim that we should rejoice in a lack of data because of the taxonomic convenience thus provided).

This traditional approach to morphological breaks merely underscores what Feyerabend meant (see above) in comparing theories to party lines, for it renders the picture of phyletic gradualism virtually unfalsifiable. The picture prescribes an interpretation and the interpretation, viewed improperly as an "objective" rendering of data, buttresses the picture. We have encountered no dearth of examples, and cite the following nearly at random. Neef (1970) encountered "apparent saltation in the *Pelicaria* lineage" (p. 464), a group of Plio-Pleistocene snails. Although he cites no lithologic or geographic data favoring either interpretation, the picture of phyletic gradualism prescribes a preference: "It is likely that the discontinuity . . . is due to a period of non-deposition. . . . The possibility that the apparent saltations in the *Pelicaria* lineage are due to the migration of advanced forms from small nearby semi-isolated populations and that deposition of the Marima Sandstone was continuous cannot be entirely excluded" (1970, p. 454).

Moreover, the picture's influence has many subtle extensions. For instance:

(1) It colors our language. We are compelled to talk of "morphological breaks" in order to be understood. But the term is not a neutral descriptor; it presupposes the truth of phyletic gradualism, for a "break" is an interruption of something continuous. (Under a deVriesian picture, for example, "breaks" are "saltations"; they are real and expressive of evolutionary processes.)

(2) It prescribes the cases that are worthy of study. If breaks are artificial, the sequences in which they abound become, *ipso facto*, poor objects for evolutionary investigation. But surely there is something insidious here: if breaks are real and stand against the picture of phyletic gradualism, then the picture itself excludes an investigation of the very cases that could place it in jeopardy.

If we doubt phyletic gradualism, we should not seek to "disprove" it "in the rocks." We should bring a new picture from elsewhere and see if it provides a more adequate interpretation of fossil evidence. In the next section, we express our doubts, display a different picture, and attempt this interpretation.

But before leaving the picture of phyletic gradualism, we wish to illustrate its pervasive influence in yet another way. Kuhn (1962) has stressed the impact of textbooks in molding the thought of new professionals. The "normal science" that they inculcate is "a strenuous and devoted attempt to force nature into the conceptual boxes supplied by professional education" (1962, p. 5).

Before the "modern synthesis" of the 1930s and '40s, English-speaking invertebrate paleontologists were raised upon two texts—Eastman's translation of Zittel (1900) and that venerable *Gray's Anatomy* of British works, Woods's *Palaeontology* (editions from 1893 to 1946, last edition reprinted five times before 1958 and still very much in use). Both present an orthodox version of phyletic gradualism. In a classic statement, ending with the sentence that serves as masthead to this section, Zittel wrote (Eastman translation, 1900, p. 10):

> Weighty evidence for the progressive evolution of organisms is afforded by fossil transitional series, of which a considerable number are known to us, notwithstanding the imperfection of the palaeontological record. By transitional series are meant a greater or lesser number of similar forms occurring through several successive horizons, and constituting a practically unbroken morphic chain. . . . With increasing abundance of palaeontological material, the more numerous and more complete are the series of intermediate forms which are brought to light.

The last edition of Woods (1946) devotes three pages to evolution; all but two paragraphs (one on ontogeny, the other on orthogenesis) to an exposition of phyletic gradualism (one page on the imperfection of the record, another on some rare examples of graded sequences).

Our current textbooks have changed the argument not at all. Moore, Lalicker, and Fischer (1952, p. 30), in listing the fossil record among "evidences of evolution," have only this to say about it: "Although lack of knowledge is immeasurably greater than knowledge, many lineages among fossils of various groups have been firmly established. These demonstrate the transformation of one species or genus into another and thus constitute documentary evidence of gradual evolution." And Easton (1960, p. 34), citing the apotheosis of our achievements, writes: "An evolutionary series represents the peak of scientific accomplishment in organizing fossil invertebrates. It purports to show an orderly progression in morphologic changes among related creatures during successive intervals of time."

That these older texts hold so strongly to phyletic gradualism should surprise no one; harder to understand is the fact that virtually all modern texts repeat the same arguments even though their warrant had disappeared, as we shall now show, with the advent of the allopatric theory of speciation.

THE BIOSPECIES AND PUNCTUATED EQUILIBRIA: A DIFFERENT PICTURE OF SPECIATION

Habits of thought in the tradition of a science are not readily changed, it is not easy to deviate from the customary channels of accumulated experience in conventionalized subjects.

—G. L. JEPSEN, 1949, p. v

An irony. The formulation of the biological species concept was a major triumph of the synthetic theory (Mayr, 1963, abridged and revised 1970, remains the indispensable source on its meaning and implications). Since paleontology has always taken its conceptual lead from biology (with practical guidance from geology), it was inevitable that paleontologists should try to discover the meaning of the biospecies for their own science.

Here we meet an ironic situation: the taxonomic perspective—one of our persistent albatrosses—dictated an approach to the biospecies. Instead of extracting its insights about evolutionary processes, we sought only its prescriptions for classification. We learned that species are populations, that they are recognized in fossils by ranges of variability, not by correspondence to idealized types. The "new systematics" ushered in the revolution in species-level classification that Darwin's theory had implied but not effected. In paleontology, its main accomplishment has been a vast condensation and elimination of spurious taxa established on typological criteria.

But the new systematics also rekindled a theoretical debate unsurpassed in the annals of paleontology for its ponderous emptiness: What is the nature of a paleontological species? In this reincarnation: can taxa designated as biospecies be recognized from fossils? Biologists insisted that the biospecies is a "real" unit of nature, a population of interacting individuals, reproductively isolated from all other groups. Yet its reality seemed to hinge upon what Mayr calls its "non-dimensional" aspect: species are distinct at any moment in time, but the boundaries between forms must blur in temporal extension—a continuous lineage cannot be broken into objective segments. Attempts to reconcile or divorce the non-dimensional biospecies and the temporal "paleospecies" creep on apace (Imbrie, 1957; Weller, 1961; McAlester, 1962; Shaw, 1969; and an entire symposium edited by Sylvester-Bradley, 1956); if obfuscation is any sign of futility, we offer the following as a plea for the termination of this discussion: "Such a plexiform lineage . . . constitutes a chronospecies (or paleospecies), and it is composed of many successional polytypic morphospecies ('holomorphospecies'), each of which is in theory the paleontological equivalent of a neontological biospecies" (Thomas, 1956, p. 24).

The discussion is futile for a very simple reason: the issue is insoluble; it is not a question of fact (phylogeny proceeds as it does no matter how we name its steps), but a debate about ways of ordering information. When Whitehead said that all philosophy was a footnote to Plato, he meant not only that Plato had identified all the major problems, but also that the problems were still debated because they could not be solved. The point is this: the hierarchical system of Linnaeus was established for his world: a world of discrete entities. It works for the living biota because most species are discrete at any moment in time. It has no objective application to evolving continua, only an arbitrary one based on subjective criteria for division. Linnaeus would not have set up the same system for our world. As Vladimir Nabokov writes in *Ada* (1969, p. 406): "Man . . . will never die, because there may never be a taxonomical point in his evolutionary progress that could be determined as the last stage of man in the cline turning him into Neohomo, or some horrible throbbing slime."

Then does the biospecies offer us nothing but semantic trouble? On one level, the answer is no because it can be applied with great effectiveness to past time-planes. But on another level, and this involves our irony, we must avoid the narrow approach that embraces a biological concept only when it can be transplanted bodily into our temporal taxonomy. The biospecies abounds with implications for the operation of evolutionary processes. Instead of attempting vainly to name successional taxa objectively in its light (McAlester, 1962), we should be applying its concepts. In the following section, we argue that one of these concepts—the theory of allopatric speciation—might reorient our picture for the origin of taxa.

Implications of allopatric speciation for the fossil record. We wish to consider an alternate picture to phyletic gradualism; it is based on a theory of speciation that arises from the behavior, ecology, and distribution of modern biospecies. First, we must emphasize that mechanisms of speciation can be studied directly only with experimental and field techniques applied to living organisms. No theory of evolutionary mechanisms can be generated directly from paleontological data. Instead, theories developed by students of the modern biota generate predictions about the course of evolution in time. With these predictions, the paleontologist can approach the fossil record and ask the following question: Are observed patterns of geographic and stratigraphic distribution, and apparent rates and directions of morphological change, consistent with the consequences of a particular theory of speciation? We can apply and test, but we cannot generate new mechanisms. If discrepancies are found between paleontological data and the expected patterns, we may be able to identify those aspects of a general theory that need improvement. But we cannot formulate these improvements ourselves.*

During the past thirty years, the allopatric theory has grown in popularity to become, for the vast majority of biologists, *the* theory of speciation. Its only serious challenger is the sympatric theory. Here we discuss only the implications of the allopatric theory for interpreting the fossil record of sexually reproducing metazoans. We do this simply because it is the allopatric, rather than the sympatric, theory that is preferred by biologists. We shall therefore contrast the allopatric theory with the picture of phyletic gradualism developed in the last section.

Most paleontologists, of course, are aware of this theory, but the influence of phyletic gradualism remains so strong that discussions of geographic speciation are almost always cast in its light: geographic speciation is seen as the slow and steady transformation of two separated lineages—i.e., as *two* cases of phyletic gradualism (FIGURE 3). Raup and Stanley (1971, p. 98), for example, write:

> Let us consider populations of a species living at a given time but not in geographic contact with each other. . . . Two or more segments of the species thus evolve and undergo *phyletic* speciation independently. . . . The distinction between phyletic and geographic speciation is to some extent artificial in that both processes depend on natural selection. The critical difference is that phyletic speciation is accomplished in the absence of geographic isolation and geographic speciation requires geographic isolation [italics ours].

The central concept of allopatric speciation is that new species can arise only when a small local population becomes isolated at the margin of the geographic range of its parent species. Such local populations are termed *peripheral isolates*. A peripheral isolate develops into a new species if *isolating mechanisms* evolve that will prevent the re-initiation of gene flow if the new form re-encounters its ancestors at some future time. As a consequence of the allopatric theory, new fossil species do not originate in the place where their ancestors lived. It is extremely improbable that we shall

* The rate and direction of morphological change over long periods of time is the most obvious kind of evolutionary pattern that we can test against predictions based on processes observed over short periods of time by neontologists. We try to do this in the next section.

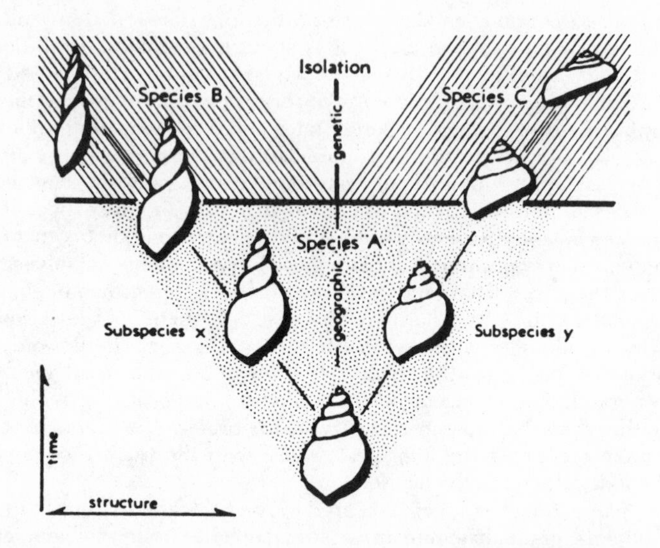

FIGURE 3: A hypothetical case of geographic speciation viewed from the perspective of phyletic gradualism—slow and gradual transformation in two lineages. From Moore, Lalicker, and Fischer, 1952; figure 1–15.

be able to trace the gradual splitting of a lineage merely by following a certain species up through a local rock column.

Another consequence of the theory of allopatric processes follows: since selection always maintains an equilibrium between populations and their local environment, the morphological features that distinguish the descendant species from its ancestor are present close after, if not actually prior to, the onset of genetic isolation. These differences are often accentuated if the two species become sympatric at a later date (character displacement—Brown and Wilson, 1956). In any event, most morphological divergence of a descendant species occurs very early in its differentiation, when the population is small and still adjusting more precisely to local conditions. After it is fully established, a descendant species is as unlikely to show gradual, progressive change as is the parental species. Thus, in the fossil record, we should not expect to find gradual divergence between two species in an ancestral–descendant relationship. Most evolutionary changes in morphology occur in a short period of time relative to the total duration of species. After the descendant is established as a full species, there will be little evolutionary change except when the two species become sympatric for the first time.

These simple consequences of the allopatric theory can be combined into an expected pattern for the fossil record. Using stratigraphic, radiometric, or biostratigraphic criteria (for organisms other than those under study), we establish a regional framework of correlation. Starting with these correlations, patterns of geographic (not stratigraphic) variation among samples of fossils should appear. Tracing a fossil species through any local rock column, so long as no drastic changes occur in the physical environment, should produce *no* pattern of constant change, but one of oscillation in mean values. Closely related (perhaps descendant) species that enter the rock column should appear suddenly and show no intergradation with the "ancestral" species in morphological features that act as inter-specific differentia. There should be no gradual divergence between the two species when both persist for some time to higher stratigraphic levels. Quite the contrary —it is likely that the two species will display their greatest difference when the descendant first appears. Finally, in exceptional circumstances, we may be able to identify the general area of the ancestor's geographic range in which the new species arose.

Another conclusion is that time and geography, as factors in evolution, are not so comparable as some authors have maintained (Sylvester-Bradley, 1951). The allopatric theory predicts that most variation will be found among samples drawn from different geographic areas rather than from different stratigraphic levels in the local rock column. The key factor is adjustment to a heterogeneous series of micro-environments vs. a general pattern of stasis through time.

In summary, we contrast the tenets and predictions of allopatric speciation with the corresponding statements of phyletic gradualism previously given:

(1) New species arise by the splitting of lineages.

(2) New species develop rapidly.

(3) A small sub-population of the ancestral form gives rise to the new species.

(4) The new species originates in a very small part of the ancestral species' geographic extent—in an isolated area at the periphery of the range.

These four statements again entail two important consequences:

(1) In any *local* section containing the ancestral species, the fossil record for the descendant's origin should consist of a sharp morphological break between the two forms. This break marks the migration of the descendant, from the peripherally isolated area in which it developed, into its ancestral range. Morphological change in the ancestor, even if directional in time, should bear no relationship to the descendant's morphology (which arose in response to local conditions in its isolated area). Since speciation occurs rapidly in small populations occupying small areas far from the center of ancestral abundance, we will rarely discover the actual event in the fossil record.

(2) Many breaks in the fossil record are real; they express the way in which evolution occurs, not the fragments of an imperfect record. The sharp break in a local column accurately records what happened in that area through time. Acceptance of this point would release us from a self-imposed status of inferiority among the evolutionary sciences. The paleontologist's gut-reaction is to view almost any anomaly as an artifact imposed by our

institutional millstone—an imperfect fossil record. But just as we now tend to view the rarity of Precambrian metazoans as a true reflection of life's history rather than a testimony to the ravages of metamorphism or the lacunae of Lipalian intervals, so also might we reassess the smaller breaks that permeate our Phanerozoic record. We suspect that this record is much better (or at least much richer in optimal cases) than tradition dictates.

Problems of phyletic gradualism. In our alternate picture of phyletic gradualism, we are not confronted with a self-contained theory from modern biology. The postulated mechanism for gradual uni-directional change is "orthoselection," usually viewed as a constant adjustment to a uni-directional change in one or more features of the physical environment. The concept of orthoselection arose as an attempt to remove the explanation of gradual morphological change from the realm of metaphysics ("orthogenesis"). It does *not* emanate from *Drosophila* laboratories, but represents a hypothetical extrapolation of selective mechanisms observed by geneticists.

Extrapolation of gradual change under selection to a complete model for the origin of species fails to recognize that speciation is primarily an ecological and geographic process. Natural selection, in the allopatric theory, involves adaptation to local conditions and the elaboration of isolating mechanisms. Phyletic gradualism is, in itself, an insufficient picture to explain the origin of diversity in the present, or any past, biota.

Although phyletic gradualism prevails as a picture for the origin of new species in paleontology, very few "classic" examples purport to document it. A few authors (MacGillavry, 1968, Eldredge, 1971) have offered a simple and literal interpretation of this situation: *in situ*, gradual, progressive evolutionary change is a rare phenomenon. But we usually explain the paucity of cases by a nearly ritualized invocation of the inadequacy of the fossil record. It *is* valid to point out the rarity of thick, undisturbed, highly fossiliferous rock sections in which one or more species occur continuously throughout the sequence. Nevertheless, if most species evolved according to the tenets of phyletic gradualism, then, no matter how discontinuous a species' occurrence in thick sections, there should be a shift in one or more variables from sample to sample up the section. This is, in fact, the situation in most cases of postulated gradualism: the "gradualism" is represented by dashed lines connecting known samples. This procedure provides an excellent example of the role of preconceived pictures in "objectively documented" cases. One of the early "classics" of phyletic gradualism, Carruthers' (1910) study of the Carboniferous rugose coral *Zaphrentites delanouei* (Milne-Edwards and Haime) and its reinterpretation by Sylvester-Bradley (1951), is of this kind. We do not say that the analysis is incorrect; the *Z. delanouei* stock may have evolved as claimed. We merely wish to show how the *a priori* picture of phyletic gradualism has imposed itself upon limited data.

How pervasive, then, is gradualism in these quasi-continuous sequences? A number of authors (including, *inter alia*, Kurtén, 1965, Mac-

Gillavry, 1968, and Eldredge, 1971) have claimed that most species show little or no change throughout their stratigraphic range. But though it is tempting to conclude that gradual, progressive morphological change is an illusion, we recognize that there is little hard evidence to support either view.

As a final, and admittedly extreme, example of *a priori* beliefs in phyletic gradualism, we cite the work of Brace (1967) on human evolution. This is all the more instructive since most paleoanthropologists, in reversing an older view that Brace still maintains, now claim that hominid evolution involves speciation by splitting as well as phyletic evolution by transformation (seen especially in the presumed coexistence of two australopithecine species in the African lower Pleistocene—Howell, 1967; Tobias, 1965; Pilbeam, 1968; Pilbeam and Simons, 1965). Brace (1967) has claimed that the fossil record of man includes four successive "stages" in direct ancestral–descendant relation. These are the Australopithecine (with two successive "phases"—the australopithecus and paranthropus), the Pithecanthropus, the Neanderthaloid, and, finally, the Modern Stage. In discussing the history of paleoanthropology, Brace shows that most denials of ancestral–descendant relationships among hominid fossils stem from a desire to avoid the conclusion that *Homo sapiens* evolved from some "lower," more "brutish" form. But Brace has lumped all such analyses under the catch phrase "hominid catastrophism." Hominid catastrophism, according to Brace, is the denial of ancestral–descendant relationships among fossils, with the invocation of extinction and subsequent migrations of new populations that arose by successive creation. Such views are, of course, absurd, but Brace would include *all* cladistic interpretations of the hominid record within "hominid catastrophism." To view hominid phylogeny as a gradual, progressive, unilineal process involving a series of stages, Brace claims, is the interpretation most consonant with evolutionary theory. His interpretation of phylogeny may be correct (though most experts deny it), but he is seriously wrong to claim that phyletic gradualism is the picture most consistent with modern biological thought. Quite apart from the issue of probable overlap in the ranges of his stages, it would be of great interest to determine the degree of stasis attained by them during any reasonably long period of time.

Application of allopatric concepts to paleontological examples. At this point, there is some justification for concluding that the picture of phyletic gradualism is poorly documented indeed, and that most analyses purporting to illustrate it directly from the fossil record are interpretations based on a preconceived idea. On the other hand, the alternative picture of stasis punctuated by episodic events of allopatric speciation rests on a few general statements in the literature and a wealth of informal data. The idea of *punctuated equilibria* is just as much a preconceived picture as that of phyletic gradualism. We readily admit our bias towards it and urge readers, in the ensuing discussion, to remember that our interpretations are as colored by our preconceptions as are the claims of the champions of phyletic gradualism by theirs. We merely reiterate: (1) that one must have some picture of speciation in mind, (2) that the data of paleontology cannot decide which picture is more adequate, and (3) that the picture of punctuated equilibria is more in accord with the process of speciation as understood by modern evolutionists.

We could cite any number of reported sequences that fare better under notions of allopatric processes than under the interpretation of phyletic gradualism that was originally applied. This is surely true for all or part of the three warhorses of the English literature: horses themselves, the Cretaceous echinoid *Micraster*, and the Jurassic oyster *Gryphaea*. Simpson (1951) has shown that the phylogeny of horses is a luxuriant, branching bush, not the ladder to one toe and big teeth that earlier authors envisioned (Matthew and Chubb, 1921). Nichols (1959) believes that *Micraster senonensis* was a migrant from elsewhere and that it did not arise and diverge gradually from *M. cortestudinarium* as Rowe (1899) had maintained. Hallam (1959, 1962) has argued that the transition from *Liostrea* to *Gryphaea* was abrupt and that *neither* genus shows *any* progressive change through the basal Liassic zones, contrary to Trueman's claim (1922, p. 258) that: "It is doubtful whether any better example of lineage of fossil forms could be found." Gould (1971b and in press) has confirmed Hallam's conclusions. Hallam interprets the sudden appearance of *Gryphaea* as the first entry into a local rock column of a species that had evolved rapidly elsewhere. He writes (1962, p. 574): "This interpretation is more in accord with the experience of most invertebrate paleontologists who, despite continued collecting all over the world and an ever-increasing amount of research, find 'cryptogenic' genera and species far more commonly than they detect gradual trends or lineages. The sort of evolution I tentatively propose for *Gryphaea* could in fact be quite normal among the invertebrates." We agree.

We choose, rather, to present two examples from our own work which we believe are interpreted best from the viewpoint of allopatric speciation. We prefer to emphasize our own work simply because we are most familiar with it and are naturally more inclined to defend our interpretations.

Gould (1969) has analyzed the evolution of *Poecilozonites bermudensis zonatus* Verrill, a pulmonate snail, during the last 300,000 years of the Bermudian Pleistocene. The specimens were collected from an alternating sequence of wind-blown sands and red soils. Formational names, dominant lithologies, and glacial-interglacial correlations are given in TABLE *1*.

The small area and striking differentiation of stratigraphic units in the Bermudian Pleistocene permit a high degree of geographic and temporal control. *P. bermudensis* (Pfeiffer) is plentiful in all post-Belmont formations; in addition, one subspecies, *P.b. bermudensis*, is extant and available for study in the laboratory.

Distinct patterns of color banding differentiate an eastern from a western population of *P. bermudensis zonatus*. The boundary between these two groups is sharp, and there are no unambiguous cases of introgression. *P. bermudensis zonatus* was divided into two stocks, evolving in parallel with little gene flow between them, throughout the entire interval of Shore Hills to Southampton time. Both eastern and western *P.b. zonatus* became extinct sometime after the deposition of Southampton dunes; they were replaced by *P.b. bermudensis,* a derivative of eastern *P.b. zonatus* which had been evolving separately in the area of St. George's Island since St. George's time. Gould (1969, 1970b) has discussed the parallel oscillation of several morphological features in both stocks of *P.b. zonatus;* these are adaptive shifts in response to glacially controlled variations in climate. Both stocks exhibit stability in other features that serve to distinguish them from their nearest relatives. There is no evidence for any gradual divergence between eastern and western *P.b. zonatus.*

Several samples of *P. bermudensis* share many features that distinguish them from *P. bermudensis zonatus.* These characters can be arranged in four categories: color, general form of the spire, thickness of the shell, and shape of the apertural lip. The ontogeny of *P.b. zonatus* illustrates the interrelation of these categories. Immature shells of *P.b. zonatus* are weakly colored, relatively wide, lack a callus, and have the lowest portion of the outer apertural lip at the umbilical border. This combination of character states is exactly repeated in the large *mature* shells of non-*zonatus* samples of *P. bermudensis.* Since every ontogenetic feature developed at or after the fifth whorl in non-*zonatus* samples is attained by whorls 3–4 in *P.b. zonatus,* Gould (1969) concludes that the non-*zonatus* samples of *P. bermudensis* are derived by paedomorphosis from *P.b. zonatus.*

These paedomorphic samples range through the entire interval of Shore Hills to Recent. The most obvious hypothesis would hold that they constitute a continuous lineage evolving separately from *P.b. zonatus.* Gould rejects this and concludes that paedomorphic offshoots arose from the *P.b. zonatus* stock at four different times; the arguments are based on details of stratigraphic and geographic distribution, as well as on morphology.

FIGURE 4 summarizes the history of splitting in the *P.b. zonatus* lineage. The earliest paedomorph, *P.b. fasolti* Gould, occurs in the Shore Hills Formation within the geographic range of eastern *P.b. zonatus. P.b. fasolti* and the contemporary population of eastern *P.b. zonatus* share a unique set of morphological features including, *inter alia,* small size at any given whorl, low spire, relatively wide shell, and a wide umbilicus. These features unite the Shore Hills paedomorph and non-paedomorph, and set them apart from all post–Shore Hills *P. bermudensis.*

In the succeeding Harrington Formation, paedomorphic samples of *P. bermudensis* lived in both the eastern and western geographic regions of *P.b. zonatus.* The eastern paedomorph, *P.b. sieglindae* Gould, may have evolved from the Shore Hills paedomorph, *P.b. fasolti.* However, both *P.b. sieglindae* and the contemporaneous population of eastern *P.b. zonatus* lack the distinctive features of all Shore Hills *P. bermudensis* and a more likely hypothesis holds that the features uniting all post–Shore Hills *P. bermudensis* were evolved only once. If this is the case, *P.b. sieglindae* is a second paedomorphic derivative of eastern *P.b. zonatus.*

TABLE *1*. STRATIGRAPHIC COLUMN OF BERMUDA.

FORMATION	DESCRIPTION	INTERPRETATION
Recent	Poorly developed brownish soil or crust	Interglacial
Southampton	Complex of eolianites and discontinuous unindurated zones	Interglacial
St. George's	Red paleosol of island-wide extent	Glacial
Spencer's Point	Intertidal marine, beach, and dune facies	Interglacial
Pembroke	Extensive eolianites and discontinuous unindurated zones	Interglacial
Harrington	Fairly continuous unindurated layer with shallow-water marine and beach facies	Interglacial
Devonshire	Intertidal marine and poorly developed dune facies	Interglacial
Shore Hills	Well-developed red paleosol of island-wide extent	Glacial
Belmont	Complex shallow-water marine, beach, and dune facies	Interglacial
Soil (?)	A reddened surface rarely seen in the Walsingham district	Glacial?
Walsingham	Highly altered eolianites	Interglacial

P.b. sieglindae differs from its contemporary paedomorph *P.b. seigmundi* Gould in that each displays the color pattern of the local non-paedomorph. Very simply, *P.b. sieglindae* is found in eastern Bermuda and shares the banding pattern of eastern *P.b. zonatus*, while *P.b. siegmundi* is found in western Bermuda and has the same color pattern as western *P.b. zonatus*. In addition, both *P.b. sieglindae* and *P.b. siegmundi* evolved at the periphery of the known range of their putative ancestors. The independent derivation of the two Harrington paedomorphs from the two stocks of *P.b. zonatus* seems clear.

Finally, the living paedomorph, *P.b. bermudensis*, first appears in the St. George's Formation on St. George's Island. While St. George's Island is within the geographic range of eastern *P.b. zonatus*, it is far removed from the area in which *P.b. sieglindae* arose and lived. Gould concludes that *P.b. sieglindae* was a short-lived population that never enjoyed a wide geographic distribution; he estimates that the Pembroke population's range did not exceed 200 meters. Although there is little morphological evidence to support it, Gould recognizes a fourth paedomorphic subspecies, *P.b. bermudensis*, derived directly from (eastern) *P.b. zonatus*. The conclusion is based upon geographic and stratigraphic data.

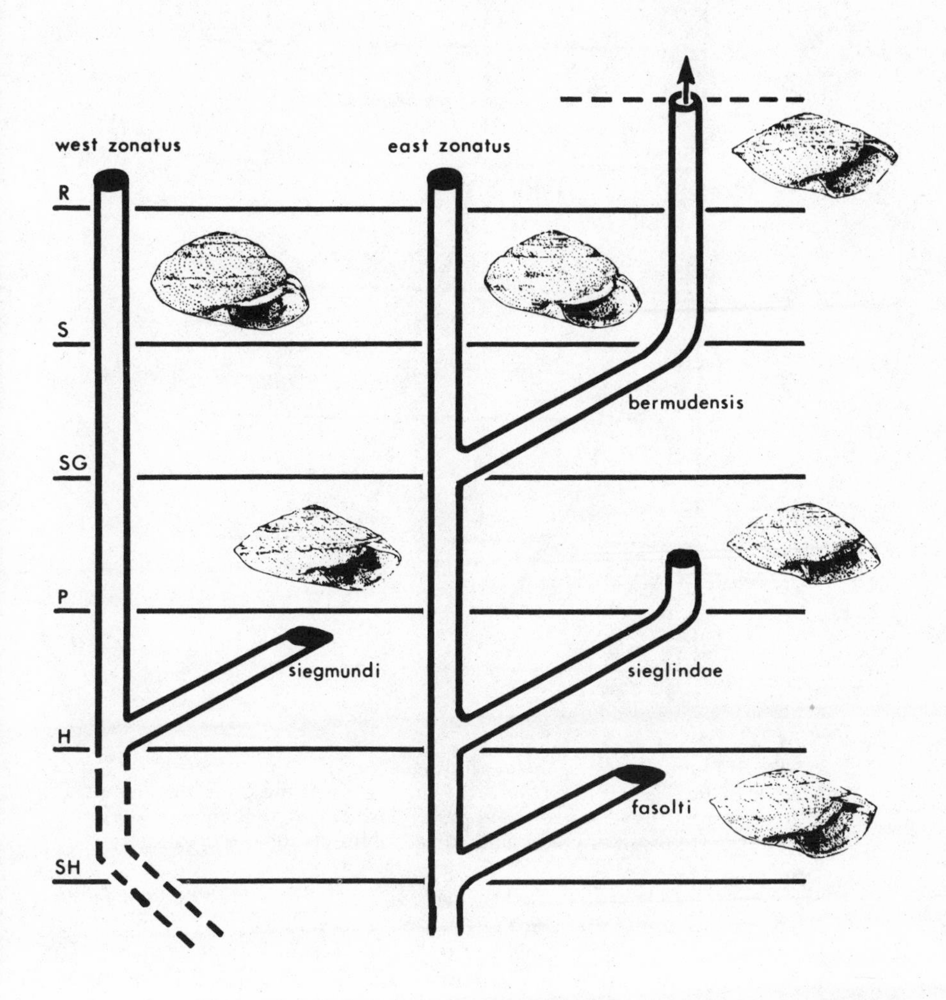

FIGURE 4: Reconstruction of the phylogenetic history of *P. bermudensis* showing iterative development of paedomorphic subspecies. SH—Shore Hills: H—Harrington; P—Pembroke; SG—St. George's; S—Southampton; R—Recent. From Gould, 1969; figure 20.

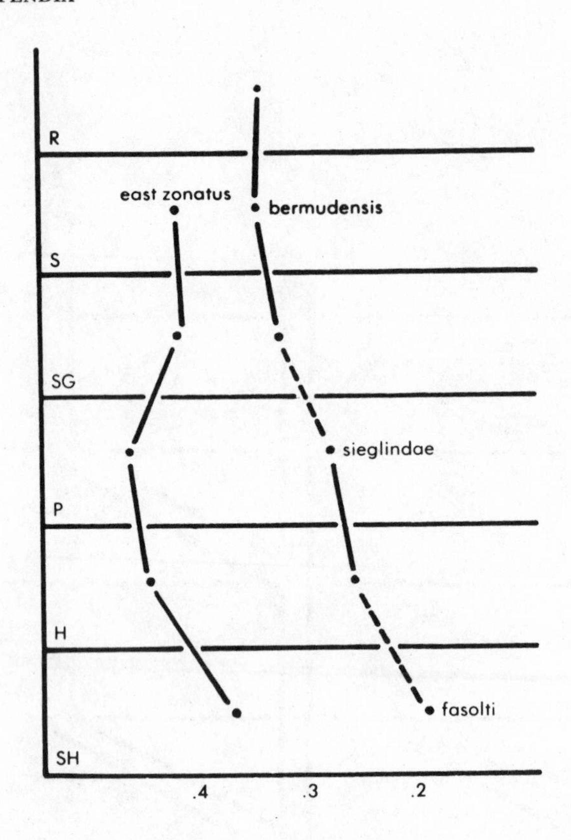

FIGURE 5: Plot of means of mean sample values of "lower eccentricity" in *P. bermudensis*. Dashed lines show the phylogeny of the three paedomorphs of eastern *zonatus* as a direct ancestral–descendant sequence, and offer a tempting instance of phyletic gradualism. Abbreviations as in FIGURE 4.

Gould (1969) has advanced an adaptive explanation for the four separate origins of paedomorphic populations from *P.b. zonatus*. This explanation, based on the value of thin shells in lime-poor soils, need not be elaborated here. What is important, for our purposes, is to emphasize that the reconstruction of phylogenetic histories for the paedomorphs involves (1) attention to geographic data (the allopatric model), (2) discontinuous stratigraphic occurrence (a more literal interpretation of the fossil record), and (3) formal arguments based on morphology. It is entirely possible, from morphological data alone, to interpret the three paedomorphs of the eastern *zonatus* area as a gradational biostratigraphic series. FIGURE 5 shows a tempting interpretation of phyletic gradualism for "lower eccentricity," an apertural variable. Values gradually increase through time. FIGURE 6, how-

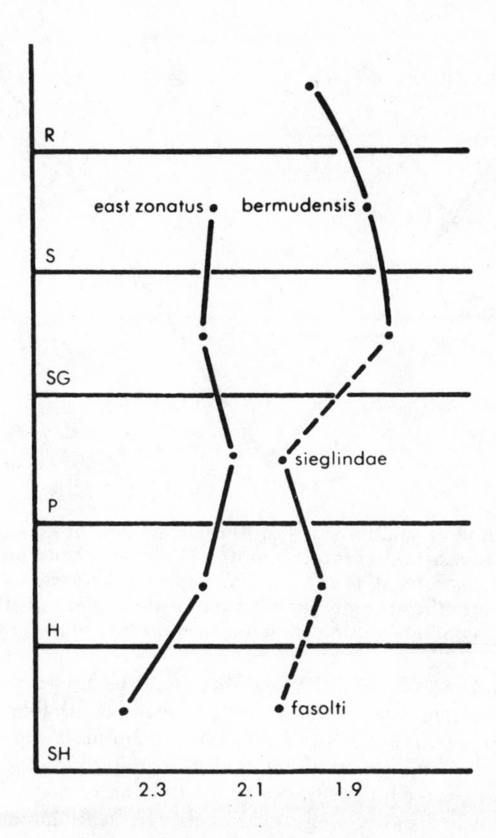

FIGURE 6: Plot of means of mean sample values for "differential growth ratio" in *P. bermudensis.* Dashed lines show the interpretation of the phylogeny of the three paedomorphs as a direct ancestral–descendant sequence. Abbreviations as in FIGURE *4*.

ever, confounds this interpretation by showing that stratigraphic variability in "differential growth ratio" within both *P.b. sieglindae* and *P.b. bermudensis* varies in a direction *opposite* to the net stratigraphic "trend": *P.b. fasolti —P.b. sieglindae—P.b. bermudensis:* This could be read to indicate that each subspecies is unique. In fact, neither graph affords sufficient evidence to warrant either conclusion. Morphology, stratigraphy, and geography must all be evaluated.

The phylogenetic history of the trilobite *Phacops rana* (Green) from the Middle Devonian of North America (Eldredge, 1971; 1972) provides another example of the postulated operation of allopatric processes. As in *Poecilozonites bermudensis,* full genetic isolation was probably not established between "parent" and "daughter" taxa; this conclusion, based on

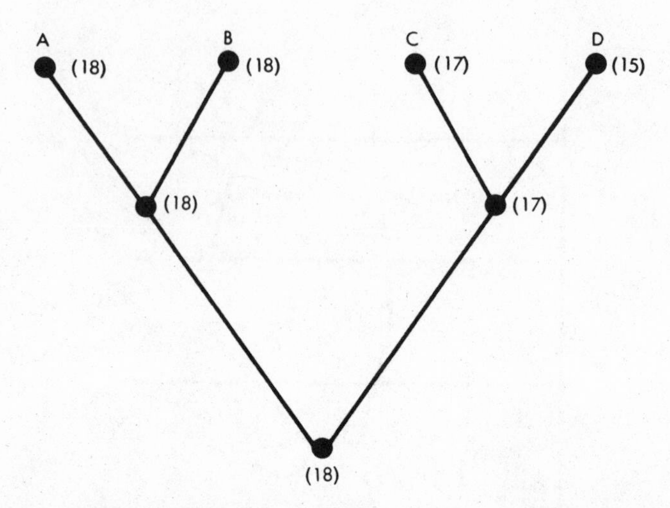

FIGURE 7: Outline of relationships of four subspecies of *Phacops rana*. A—*Phacops rana crassituberculata* Stumm; B—*Phacops rana milleri* Stewart: C—*Phacops rana rana* (Green): D—*Phacops rana norwoodensis* Stumm. Numbers in parentheses refer to number of dorso-ventral files typical of subspecies or hypothesized to characterize condition of common ancestor.

inferences from morphological variability, may be unwarranted. For our purposes, it does not matter whether we are dealing with four subspecies of *P. rana,* or four separate species of *Phacops,* including *P. rana* and its three closest relatives. The basic mode of evolution underlying the group's phylogenetic history as a whole is the same in either case.

Features of eye morphology exhibit the greatest amount of variation among samples of *P. rana*. Lenses are arranged on the visual surface of the eye in vertical dorso-ventral files (Clarkson, 1966). A stable number of dorso-ventral files, characteristic of the entire sample in any population, is reached early in ontogeny. The number of dorso-ventral (d.-v.) files is the most important feature of interpopulational variation in *P. rana*.

The closest known relative of *P. rana* is *P. schlotheimi* (Bronn) *s.l.*, from the Eifelian of Europe and Africa; this group has recently been revised by C. J. Burton (1969). In addition, several samples of *P. rana* have been found in the Spanish Sahara in northwestern Africa (Burton and Eldredge, in preparation). *P. schlotheimi* and the African specimens of *P. rana* are most similar to *P. rana milleri* Stewart and *P. rana crassituberculata* Stumm, the two oldest subspecies of *P. rana* in North America. All these taxa possess 18 dorso-ventral files. Eldredge (1972) concludes that 18 is the primitive number of d.-v. files for all North American *Phacops rana*.

FIGURE 7 summarizes relationships among the four subspecies of *P. rana* without regard to stratigraphic occurrence. The oldest North American *P. rana* occurs in the Lower Cazenovian Stage of Ohio and central New York

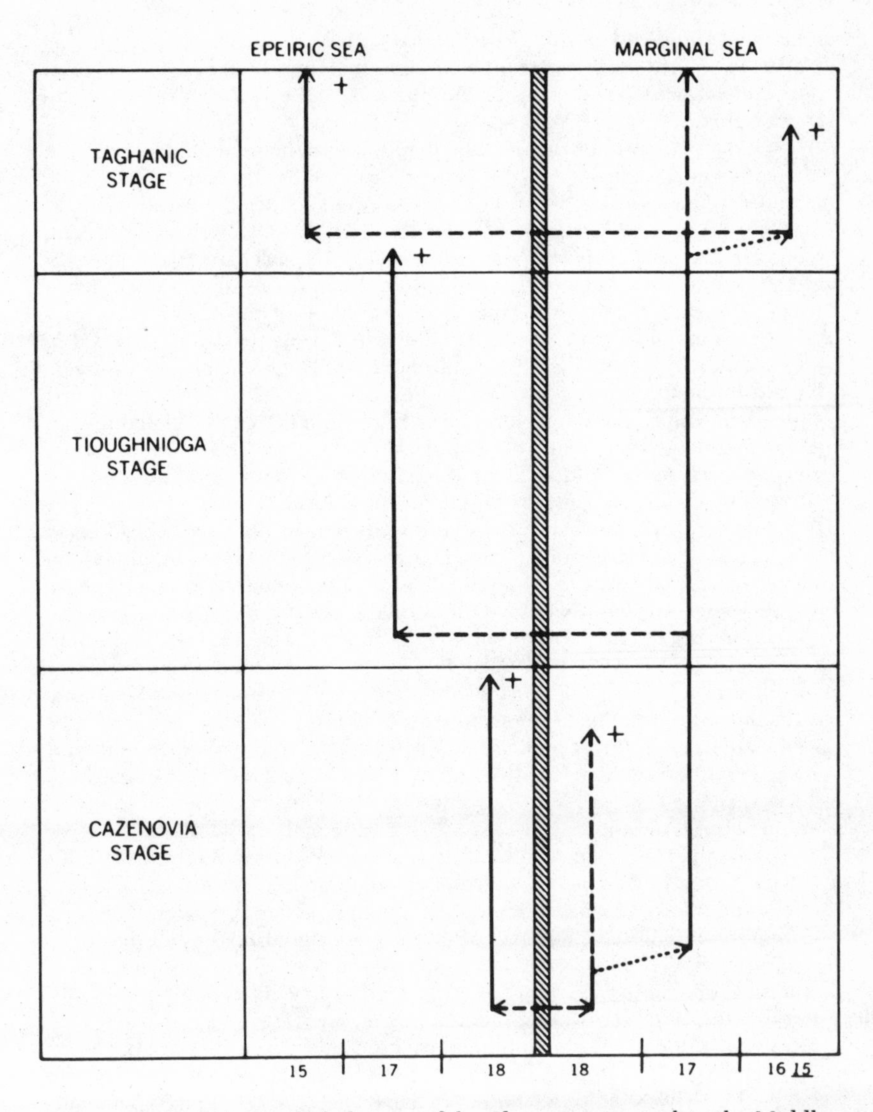

FIGURE 8: Hypothesized phylogeny of the *Phacops rana* stock in the Middle Devonian of North America. Numbers at the base of the diagram refer to the population number of dorso-ventral files. Dotted lines: origin of new (reduced) number of d.-v. files in a peripheral isolate; horizontal dashed lines: migration; vertical solid lines: presence of taxon in indicated area; dashed vertical lines: persistence of ancestral stock in a portion of the marginal sea other than that in which the derived taxon occurs. Crosses denote final disappearance; for fuller explanation, see text.

State. All have 18 d.-v. files. Populations with 18 d.-v. files *(P. rana milleri* and *P. rana crassituberculata)* persist into the Upper Cazenovian Stage in the epicontinental seas west of the marginal basin in New York and the Appalachians.

Of the two samples the one that displays intra-populational variation in d.-v. file number occurs in the Lower Cazenovian of central New York. Some specimens have 18 d.-v. files, while others reduce the first d.-v. file to various degrees; a few lack it altogether. *All P. rana* from subsequent, younger horizons in New York and adjacent Appalachian states have 17 dorso-ventral files. Apparently, 17 d.-v. file *P. rana rana* arose from an 18 d.-v. file population on the northeastern periphery of the Cazenovian geographic range of *P. rana*. Seventeen d.-v. file *P. rana* persist, unchanged in most respects, through the Upper Cazenovian, Tioughniogan, and Taghanic Stages in the eastern marginal basin. Seventeen d.-v. file *P. rana rana* first appears in the shallow interior seas at the beginning of the Tioughniogan Stage, replacing the 18 d.-v. file populations that apparently became extinct during a general withdrawal of seas from the continental interior. All Tioughniogan *P. rana* possess 17 dorso-ventral files.

A second, similar event involving reduction in dorso-ventral files occurred during the Taghanic. Here again, a variable population inhabited the eastern marginal basin in New York. This suggests that, once more, reduction in d.-v. files occurred allopatrically on the periphery of the known range of *P. rana rana*. The subsequent spread of stabilized, 15 d.-v file *P. rana norwoodensis* through the Taghanic seas of the continental interior was instantaneous in terms of our biostratigraphic resolution. FIGURE *8* summarizes this interpretation of the history of *P. rana*.

Under the tenets of phyletic gradualism, this story has a different (and incorrect) interpretation: the three successional taxa of the epeiric seas form an *in situ* sequence of gradual evolutionary modification. The sudden transitions from one form to the next are the artifact of a woefully incomplete fossil record. Most evolutionary change occurred during these missing intervals: fill in the lost pieces with an even dotted line.

If the interpreter pays attention to geographic detail, however, quite a different tale emerges, one that allows a more literal reading of the fossil record. Now the story is one of stasis: no variation in the most important feature of discrimination (number of d.-v. files—actually a complex of highly interrelated variables) through long spans of time. Two samples displaying intra-populational variation in numbers of d.-v. files identify relatively "sudden" events of reduction in files on the periphery of the species' geographic range. These two samples, moreover, have a very short stratigraphic, and very restricted geographic, distribution.

Our two examples, so widely separated in scale, age, and subject, have much in common as exemplars of allopatric processes. Both required an attention to details of *geographic* distribution for their elucidation. Both involved a *more literal* reading of the fossil record than is allowed under the unconscious guidance of phyletic gradualism. Both are characterized by *rapid* evolutionary events punctuating a history of stasis. These are among the expected consequences if most fossil species arose by allopatric speciation in small, peripherally isolated populations. This alternative picture merely represents the application to the fossil record of the dominant theory of speciation in modern evolutionary thought. We believe that the consequences of this theory are more nearly demonstrated than those of phyletic gradualism by the fossil record of the vast majority of Metazoa.

SOME EXTRAPOLATIONS TO MACROEVOLUTION

Before 1930, paleontology sought a separate theory for the causes of macroevolution. The processes of microevolution (including the origin of species) were deemed insufficient to generate the complexity and diversity of life, even under the generous constraint of geological time; a variety of special causes were proposed—vitalism, orthogenesis, racial "life" cycles, and universal acceleration in development to name just a few.

However, the advent of the "modern synthesis" inspired a reassessment that must stand as the major conceptual advance in 20th-century paleontology. Special explanations for macroevolution were abandoned for a simplifying theory of extrapolation from species-level processes. All evolutionary events, including those that seemed most strongly "directed" and greatly extended in time, were explained as consequences of mutation, recombination, selection, etc.—i.e., as consequences only of the phenomena that produce evolution in nature's real taxon, the species. (The modern synthesis received its name because it gathered under one theory—with population genetics at its core—the events in many subfields that had previously been explained by special theories unique to that discipline. Such an occurrence marks scientific "progress" in its truest sense—the replacement of special explanations carrying little power in prediction or extension with general theories, rich in implications and capable of unifying a diverse set of phenomena that had seemed unrelated. Thus Simpson (1944, 1953) did for paleontology what Dobzhansky (1937) had done for classical genetics, Mayr (1942) for systematics, de Beer (1940) for development, White (1954) for cytology, and Stebbins (1950) for systematic botany—he exemplified the phenomena of his field as the result of Darwinian processes acting upon species.)

We have discussed two pictures for the origin of species in paleontology. In the perspective of a species-extrapolation theory of macroevolution, we should now extend these pictures to see how macroevolution proceeds under their guidance. If actual events, as recorded by fossils, fit more comfortably with the predictions of either picture, this will be a further argument for that picture's greater adequacy.

Under phyletic gradualism, the history of life should be one of *stately unfolding*. Most changes occur slowly and evenly by phyletic transformation; splitting, when it occurs, produces a slow and very gradual divergence

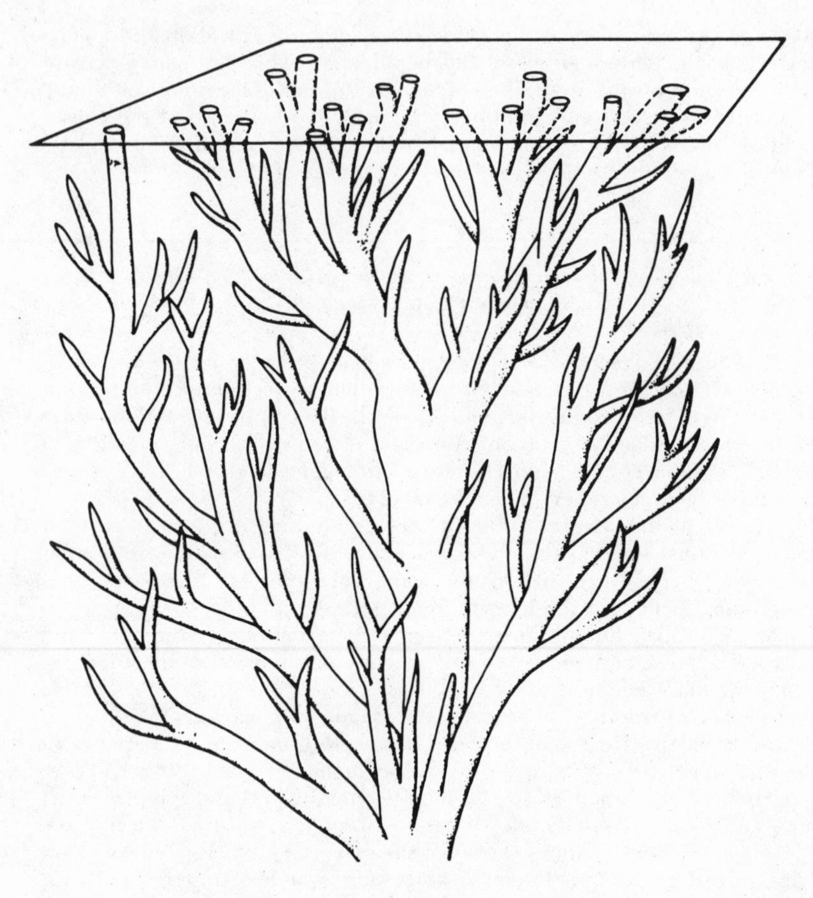

FIGURE 9: The "Tree of Life" viewed from the perspective of phyletic gradualism. Branches diverge gradually one from the other. A slow and relatively equal rate of evolution pervades the system. From Weller, 1969; figure 637.

of forms (Weller's [1969] tree of life—reproduced as FIGURE 9—records the extrapolation of this partisan view, not a neutral hatrack for the fossils themselves). We have already named our alternate picture for its predicted extrapolation—*punctuated equilibria*. The theory of allopatric speciation implies that a lineage's history includes long periods of morphologic stability, punctuated here and there by rapid events of speciation in isolated subpopulations.

We now consider two phenomena of macroevolution as case studies of our extrapolated pictures. The first is widely recognized as anomalous under the unconscious guidance of stately unfolding; it emerges as an expectation under the notion of punctuated equilibria. The second phenomenon seems, superficially, to have an easier explanation under stately unfolding, but we shall argue that is has a more interesting interpretation when viewed with the picture of punctuated equilibria.

(1) *"Classes" of great number and low diversity*

To many paleontologists, nothing is more distressing than the current situation in echinoderm systematics. Ubaghs (1967), in his contribution to the *Treatise on Invertebrate Paleontology*, recognizes 20 classes and at least one has been added since then—Robison and Sprinkle's (1969) ctenocystoids. Yet, although all appeared by the Ordovician, only five survived the Devonian. Moreover, although each class has a distinct Bauplan, many display a diversity often considered embarrassingly small for so exalted a taxonomic rank—the *Treatise* describes eight classes with five or fewer genera; five of these include but a single genus (as does the new ctenocystoids).

There are two aspects to this tale that fit poorly with the traditional view of stately unfolding:

(1) The presence of 21 classes by the Ordovician, coupled with their presumed monophyletic descent, requires extrapolation to a common ancestor uncomfortably far back in the Precambrian if Ordovician diversity is the apex of a gradual unfolding. Yet current views of Precambrian evolution will not happily accommodate a complex metazoan so early (Cloud, 1968).

(2) We expect that successively higher ranks of the taxonomic hierarchy will contain more and more taxa: a class with one genus is anomalous and we are led either to desperate hopes for synonymy or, once again, to our old assumption—that we possess a fragmentary record of a truly diverse group. Yet this expectation is no consequence of the logic of taxonomy (which demands only that each taxon be *as* inclusive as the lower ones it incorporates); it arises, rather, from a picture of stately unfolding. In FIGURE 9, a new higher taxon attains its rank by *virtue of* its diversity—an evenly progressing, evenly diverging set of branches cannot produce such a taxon with limited diversity, for a lineage "graduates" from family to order to class only as it persists to a tolerable age and branches an acceptable number of times.

With the picture of punctuated equilibria, however, classes of small membership are welcome and echinoderm evolution becomes more intriguing than bothersome. Since speciation is rapid and episodic, repeated splitting during short intervals is likely when opportunities for full speciation following isolation are good (limited dangers of predation or competition in peripheral environments, for example—a likely Lower Cambrian situation). When these repeated splits affect a small, isolated lineage; when adaptation to peripheral environments involves new modes of feeding, protection, and locomotion; and when extinction of parental species commonly follows the migration of descendants to the ancestral area, then very distinct phenons with few species will develop. Since higher taxa are all "arbitrary" (they reflect no interacting group in nature, but rather a convenient arrangement of species that violates no rule of monophyly, hierarchical ordering, etc.), we believe that they should be defined by morphology. Criteria of diversity are too closely tied to partisan pictures; morphology, though not as "objective" as some numerical taxonomists claim, is at least more functional for information retrieval.

(2) *Trends*

Trends, or biostratigraphic character gradients, are frequently mentioned as basic features of the fossil record. Sequences of fossils, said to display trends, range from the infraspecific through the very highest levels of the taxonomic hierarchy. Trends at and below the species level were discussed in the previous section, but the relation between phyletic gradualism and trends among related clusters of species—families or orders—remains to be examined.

Many, if not most, trends involving higher taxa may simply reflect a selective rendering of elements in the fossil record, chosen because they seem to form a morphologically graded series coincident with a progressive biostratigraphic distribution. In this sense, trends may represent simple extrapolations of phyletic gradualism.

But a claim that all documented trends are just unwarranted extrapolations based on a preconception would be altogether too facile an explanation for the large numbers of trends cited in the literature. For this discussion, we accept trends as a real and important phenomenon in evolution, and adopt the simple definition given by MacGillavry (1968, p. 72: "A trend is a direction which involves the *majority* of related lineages of a group" (our italics).

If trends are real and common, how can they be reconciled with our picture, in which speciation occurs in peripheral isolates by adaptation to local conditions and the perfection of isolating mechanisms? The problem may be stated in another way: Sewall Wright (1967, p. 120) has suggested that, just as mutations are stochastic with respect to selection within a population, so might speciation be stochastic with respect to the origin of higher taxa. As a slight extension of that statement, we might claim that adaptations to local conditions by peripheral isolates are stochastic with respect to long-term net directional change (trends) within a higher taxon as a whole. We are left with a bit of a paradox: to picture speciation as an allopatric phenomenon, involving rapid differentiation within a general, long-term picture of stasis, is to deny the picture of directed gradualism in speciation. Yet, superficially at least, this directed gradualism is easier to reconcile with valid cases of long-term trends involving many species.

MacGillavry's definition of a trend removes part of the problem by using the expression "majority of related lineages." This frees us from the constraint of reconciling *all* events of adaptation to local conditions in peripheral isolates, with long-term, net directional change.

A reconciliation of allopatric speciation with long-term trends can be formulated along the following lines: we envision multiple "explorations" or "experimentations" (see Schaeffer, 1965)—i.e., invasions, on a stochastic basis, of new environments by peripheral isolates. There is nothing inherently directional about these invasions. However, a subset of these new environments might, in the context of inherited genetic constitution in the ancestral components of a lineage, lead to new and improved efficiency.

Improvement would be consistently greater within this hypothetical subset of local conditions that a population might invade. The overall effect would then be one of net, apparently directional change: but, as in the case of selection upon mutations, the initial variations would be stochastic with respect to this change (FIGURE *10*). We postulate no "new" type of selection. We simply state a view of long-term, superficially "directed" phenomena that is in accord with the theory of allopatric speciation, and also avoids the largely untestable concept of orthoselection.

CONCLUSION: EVOLUTION, STATELY OR EPISODIC?

Heretofore, we have spoken of the morphological stability of species in time without examining the reasons for it. The standard definition of a biospecies—as a group of actually or potentially reproducing organisms sharing a common gene pool—specifies the major reason usually cited: gene flow. Since the subpopulations of a species adapt to a range of differing local environments, we might expect these groups to differentiate, acquire isolating mechanisms and, eventually, to form new species. But gene flow exerts a homogenizing influence "to counteract local ecotypic adaptation by breaking up well-integrated gene complexes" (Mayr, 1963, p. 178). The role of gene flow is recognized in the central tenet of allopatric speciation: speciation occurs in *peripheral* isolates because only geographic separation from the parental species can reduce gene flow sufficiently to allow local differentiation to proceed to full speciation.

Recently, however, a serious challenge to the importance of gene flow in species' cohesion has come from several sources (Ehrlich and Raven, 1969, for example). Critics claim that, in most cases, gene flow is simply too restricted to exert a homogenizing influence and prevent differentiation. This produces a paradox: why, then, are species coherent (or even recognizable)? Why do groups of (relatively independent) local populations continue to display a fairly consistent phenotype that permits their recognition as a species? Why does reproductive isolation not arise in every local population? Why is the local population itself not considered the "real" unit in evolution (as some would prefer—Sokal and Crovello, 1970, p. 151, for example)?

The answer probably lies in a view of species and individuals as homeostatic systems—as amazingly well buffered to resist change and maintain stability in the face of disturbing influences. This concept has been urged particularly by Lerner (1954) and Mayr (1963), though the latter still gives more weight to gene flow than many will allow. Lerner (1954, p. 6) recognizes two types of homeostasis, mediated in both cases, he believes, by the generally higher fitness of heterozygous vs. homozygous genotypes: (1) ontogenetic self-regulation (developmental homeostasis) "based on the greater ability of the heterozygote to stay within the norms of canalized

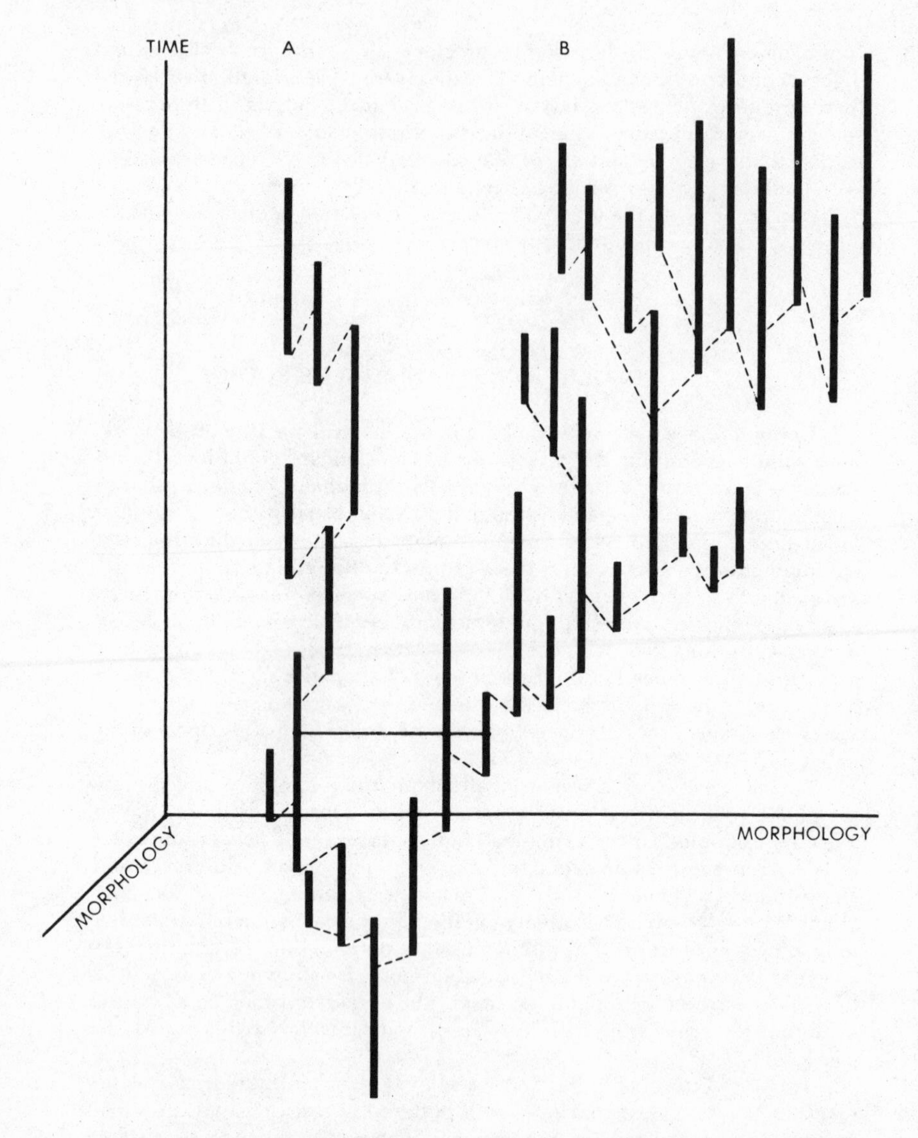

FIGURE *10*: Three-dimensional sketch contrasting a pattern of relative stability (A) with a trend (B), where speciation (dashed lines) is occurring in both major lineages. Morphological change is depicted here along the horizontal axes, while the vertical axis is time. Though a retrospective pattern of directional selection might be fitted as a straight line in (B), the actual pattern is stasis within species, and differential success of species exhibiting morphological change in a particular direction. For further explanation, see text.

development" and (2) self-regulation of populations (genetic homeostasis) "based on natural selection favoring intermediate rather than extreme phenotypes." In this view, the importance of peripheral isolates lies in their small size and the alien environment beyond the species border that they inhabit—for only here are selective pressures strong enough and the inertia of large numbers sufficiently reduced to produce the "genetic revolution" (Mayr, 1963, p. 533) that overcomes homeostasis. The coherence of a species, therefore, is not maintained by interaction among its members (gene flow). It emerges, rather, as a historical consequence of the species' origin as a peripherally isolated population that acquired its own powerful homeostatic system. (We regard this idea as a serious challenge to the conventional view of species' reality that depends upon the organization of species as ecological units of *interacting* individuals in nature. If groups of nearly independent local populations are recognized as species only because they share a set of homeostatic mechanisms developed long ago in a peripheral isolate that was "real" in our conventional sense of interaction, then some persistent anomalies are resolved. The arrangement of many asexual groups into good phenetic "species," quite inexplicable if interaction is the basis for coherence, receives a comfortable explanation under notions of homeostasis.)

Thus, the challenge to gene flow that seemed to question the stability of species in time ends by reinforcing that stability even more strongly. If we view a species as a set of subpopulations, all ready and able to differentiate but held in check only by the rein of gene flow, then the stability of species is a tenuous thing indeed. But if that stability is an inherent property of both individual development and the genetic structure of populations, then its power is immeasurably enhanced, for the basic property of homeostatic systems, or steady states, is that they resist change by self-regulation. That local populations do not differentiate into species, even though no external bar prevents it, stands as strong testimony to the inherent stability of species in time.

Paleontologists should recognize that much of their thought is conditioned by a peculiar perspective that they must bring to the study of life: they must look down from its present complexity and diversity into the past; their view must be retrospective. From this vantage point, it is very difficult to view evolution as anything but an easy and inevitable result of mere existence, as something that unfolds in a natural and orderly fashion. Yet we urge a different view. The norm for a species or, by extension, a community is stability. Speciation is a rare and difficult event that punctuates a system in homeostatic equilibrium. That so uncommon an event should have produced such a wondrous array of living and fossil forms can only give strength to an old idea: paleontology deals with a phenomenon that belongs to it alone among the evolutionary sciences and that enlightens all its conclusions—time.

BIBLIOGRAPHY

Anonymous, 1967. *Did Man Get Here by Evolution or by Creation?* Watchtower Bible and Tract Society of Pennsylvania, New York.

de Beer, G., 1940. *Embryos and Ancestors*. Oxford University Press, Oxford. 3rd ed. 1958.

———1970. "The Evolution of Charles Darwin," review of M. T. Ghiselin, *The Triumph of the Darwinian Method*. *New York Review of Books* 15:31–35.

Brace, C. L., 1967. *The Stages of Human Evolution*. Prentice-Hall, Englewood Cliffs, N.J.

Brown, W. L., Jr., and E. O. Wilson, 1956. "Character displacement." *Systematic Zoology 14:* 101–9.

Burton, C. J., 1969. "Variation studies of some phacopid trilobites of Eurasia and North West Africa." Doctoral dissertation. University of Exeter, Exeter, England.

Cain, A. J., 1954. *Animal Species and their Evolution*. Hutchinson and Co., Ltd.

Carruthers, R. G., 1910. "On the evolution of *Zaphrentis delanouei* in Lower Carboniferous times." *Quarterly Journal of the Geological Society of London* 66:523–38.

Clarkson, E. N. K., 1966. "Schizochroal eyes and vision of some Silurian acastid trilobites." *Palaeontology* 9:1–29.

Cloud, P. E., Jr., 1968. "Pre-metazoan evolution and the origins of the Metazoa," in *Evolution and Ecology*, E. T. Drake (ed.). Yale University Press, New Haven, Conn. pp. 1–72.

Coope, G. R., 1979. "Late Cenozoic fossil Coleoptera: Evolution, biogeography and ecology." *Annual Review of Ecology and Systematics* 10:247–67.

Darwin, C., 1859. *On the Origin of Species by Means of Natural Selection, or the Preservation of Favoured Races in the Struggle for Life*. John Murray, London. Facsimile ed., 1967, Atheneum, New York.

Dobzhansky, T., 1937. *Genetics and the Origin of Species*. Columbia University Press, New York. 2nd ed., 1941; 3rd ed., 1951; reprinted ed., 1982.

Easton, W. H., 1960. *Invertebrate Paleontology*. Harper & Row, New York.

Eaton, T. H., Jr., 1970. *Evolution*. W. W. Norton, New York.

Ehrlich, P. R., and P. H. Raven, 1969. "Differentiation of populations." *Science* 165:1228–32.

Eldredge, N., 1971. "The allopatric model and phylogeny in Paleozoic invertebrates." *Evolution* 25:156–67.

———1972. "Systematics and evolution of *Phacops rana* (Green, 1832) and *Phacops iowensis* Delo, 1935 (Trilobita) in the Middle Devonian of North America." *Bulletin of The American Museum of Natural History* 47:45–114.

———1982. *The Monkey Business: A Scientist Looks at Creationism*. Washington Square Press (Pocket Books), New York.

———1985. *Unfinished Synthesis*. Oxford University Press, New York.

———and J. Cracraft, 1980. *Phylogenetic Patterns and the Evolutionary Process*. Columbia University Press, N.Y.

———and S. J. Gould, 1972. "Punctuated equilibria: an alternative to phyletic gradualism," in T. J. M. Schopf, (ed.), *Models in Paleobiology*, pp. 82–115. Freeman, Cooper and Co., San Francisco.

———1974. Reply to Hecht. *Evolutionary Biology* 7:303–8.

———and I. Tattersall, 1982. *The Myths of Human Evolution*. Columbia University Press, New York.

Feyerabend, P. K., 1970. "Classical empiricism," in *The Methodological Heritage of Newton*, R. E. Butts and J. W. Davis (eds.). University of Toronto Press, Toronto, pp. 150–70.

Ghiselin, M. T., 1969. *The Triumph of the Darwinian Method*. University of California Press, Berkeley and Los Angeles.

———1974. "A radical solution to the species problem." *Systematic Zoology* 23:536–44.

Goldschmidt, R., 1940. *The Material Basis of Evolution*. Yale University Press, New Haven. Reprint ed., 1982.

Gould, S. J., 1969. "An evolutionary microcosm: Pleistocene and Recent history of the land snail *P. (Poecilozonites)* in Bermuda." *Bulletin of the Museum of Comparative Zoology* 138:407–531.

———1970a. "Evolutionary paleontology and the science of form." *Earth-Science Reviews* 6:77–119.

———1970b. "Coincidence of climatic and faunal fluctuations in Pleistocene Bermuda." *Science* 168:572–73.

———, 1971a. "D'Arcy Thompson and the science of form." *New Literary History* 2:229–58.

———, 1971b. "Geometric similarity in allometric growth: a contribution to the problem of scaling in the evolution of size." *American Naturalist* 105:113–36.

———, 1972. "Allometric fallacies and the evolution of *Gryphaea:* a new interpretation based on White's criterion of geometric similarity." *Evolutionary Biology* vol. 6:91–119.

———and N. Eldredge, 1977. "Punctuated equilibria: the tempo and mode of evolution reconsidered." *Paleobiology* 3:115–51.

Hallam, A., 1959. "On the supposed evolution of *Gryphaea* in the Lias." *Geological Magazine* 96:99–108.

———, 1962. "The evolution of *Gryphaea*." *Geological Magazine* 99:571–74.

———, 1973. *A Revolution in the Earth Sciences.* Clarendon Press, Oxford.

Hanson, N. R., 1969. *Perception and Discovery: An Introduction to Scientific Inquiry.* Freeman, Cooper and Co., San Francisco.

———, 1970. "Hypotheses fingo," in *The Methodological Heritage of Newton,* R. E. Butts and J. W. Davis (eds.). University of Toronto Press, Toronto, pp. 14–33.

Howell, F. C., 1967. "Recent advances in human evolutionary studies." *Quarterly Review of Biology* 42:471–513.

Hull, D. L., 1976. "Are species really individuals?" *Systematic Zoology* 25:174–91.

Imbrie, J., 1957. "The species problem with fossil animals," in *The Species Problem,* E. Mayr (ed.). American Association for the Advancement of Science, Publication No. 50, pp. 125–53.

Jepsen, G. L., E. Mayr, and G. G. Simpson, (eds.), 1963. *Genetics, Paleontology and Evolution.* Princeton University Press, Princeton, N.J.

Koyré, A., 1968. *Newtonian Studies.* University of Chicago Press, Chicago.

Kuhn, T. S., 1962. *The Structure of Scientific Revolutions.* University of Chicago Press, Chicago; 2nd ed. 1970.

Kurtén, B., 1965. "Evolution in geological time," in *Ideas in Modern Biology,* J. A. Moore (ed.). Natural History Press, Garden City, New York, pp. 329–54.

Leopold, L. B., 1969. "Quantitative comparison of some aesthetic factors among rivers." *U. A. Geological Survey,* Cicrular No. 620.

Lerner, I. M., 1954. *Genetic Homeostasis.* John Wiley, New York.

MacGillavry, H. J., 1968. "Modes of evolution mainly among marine invertebrates." *Bijdragen tot de dierkunde* 38:69–74.

Mach, E., 1893. *The Science of Mechanics.* Open Court Publishing Co., La Salle, Ill.; trans. by T. J. McCormack from 2nd ed.

Mandelbaum, M., 1964. *Philosophy, Science and Sense Perception: Historical and Critical Studies.* The Johns Hopkins Press, Baltimore.

Matthew, W. D., 1915. "Climate and evolution." *Annals of the New York Academy of Sciences* 24:171–318.

Mayr, E., 1942. *Systematics and the Origin of Species.* Columbia University Press, New York. Reprint ed., 1982.

———, 1959. "Isolation as an evolutionary factor." *Proceedings of the American Philosophical Society* 103:221–30.

———, 1963. *Animal Species and Evolution.* Harvard University Press, Cambridge, Mass.

———, 1970. *Populations, Species and Evolution.* Harvard University Press, Cambridge, Mass.

———and W. B. Provine (eds.), 1980. *The Evolutionary Synthesis: Perspec-*

tives on the Unification of Biology. Harvard University Press, Cambridge, Mass.

McAlester, A. L., 1962. "Some comments on the species problem." *Journal of Paleontology* 36:1377–81.

Medawar, P. B., 1969. *Induction and Intuition in Scientific Thought*. American Philosophical Society, Philadelphia.

Merton, R. K., 1965. *On the Shoulders of Giants*. Harcourt, Brace and World, New York.

Moore, R. C., C. G. Lalicker, and A. G. Fischer, 1952. *Invertebrate Fossils*. McGraw-Hill, New York.

Nabokov, V. 1969. *Ada or Ardor: A Family Chronicle*. McGraw-Hill, New York.

Neef, G., 1970. "Notes on the subgenus *Pelicaria*." *New Zealand Journal of Geology and Geophysics* 13:436–76.

Nichols, D., 1959. "Changes in the chalk heart-urchin *Micraster* interpreted in relation to living forms." *Philosophical Transactions of the Royal Society of London*, Series B. 242:347–437.

Penny, D., 1983. "Charles Darwin, gradualism and punctuated equilibrium." *Systematic Zoology* 32:72–4.

Pilbeam, D. R., 1968. Human Origins. *Advancement of Science* 24: 368–78.

———and Simons, E. L., 1965. "Some problems of hominid classification." *American Scientist* 53:237–59.

Popper, K. R., 1934 (English ed., 1959). *The Logic of Scientific Discovery*. Harper and Row, New York.

Raup, D. M., and S. M. Stanley, 1971. *Principles of Paleontology*. W. H. Freeman and Co., San Francisco.

Rhodes, F. H. T., 1983. "Gradualism, punctuated equilibrium and the *Origin of Species*." *Nature* 305:269–72.

Rifkin, J., 1983. *Algeny*. The Viking Press, New York.

Rightmire, G. P., 1981. "Patterns in the evolution of *Homo erectus*." *Paleobiology* 7:241–6.

Robison, R. A., and J. Sprinkle, 1969. "Ctenocystoidea: new class of primitive echinoderms." *Science* 166:1512–14.

Rowe, A. W., 1899. "An analysis of the genus *Micraster*, as determined by rigid zonal collecting from the zone of *Rhynchonella Cuvieri* to that of *Micraster cor-anguinum*." *Quarterly Journal of the Geological Society of London* 55:494–547.

Ruse, M., 1982. *Darwinism Defended*. Addison-Wesley, Reading, Mass.

Schaeffer, B., 1965. "The Role of experimentation in the origin of higher levels of organization." *Systematic Zoology* 14:318–36.

Shaw, A. B., 1969. "Adam and Eve, paleontology and the non-objective arts." *Journal of Paleontology* 43:1085–98.

Simpson, G. G., 1944. *Tempo and Mode in Evolution*. Columbia University Press, New York. Reprint ed., 1984.

———, 1951. *Horses*. Oxford University Press, New York.

———, 1953. *The Major Features of Evolution*. Columbia University Press, New York.

———, 1961. *Principles of Animal Taxonomy*. Columbia University Press, New York.

Sokal, R. R., and T. J. Crovello, 1970. "The biological species concept: a critical evaluation." *American Naturalist* 104:127–53.

Stebbins, G. L., 1950. *Variation and Evolution in Plants*. Columbia University Press, New York.

Sylvester-Bradley, P. C., 1951. "The subspecies in paleontology." *Geological Magazine* 88:88–102.

——— (ed.), 1956. *The Species Concept in Paleontology,* The Systematics Association Publication No. 2. Systematics Association, London.

Teggart, F. J., 1925. *Theory of History*. Yale University Press, New Haven. Reprinted 1977 (as *Theory and Processes of History*), University of California Press, Berkeley.

Thomas, G., 1956. "The species conflict," *ibid.*, pp. 17–31.

Thompson, D'A. W., 1942. *On Growth and Form,* 2nd ed. Cambridge University Press, Cambridge, Mass.

Tobias, P. V., 1965. "Early man in East Africa." *Science* 149:22–23.

Trueman, A. E., 1922. "The use of *Gryphaea* in the correlation of the Lower Lias." *Geological Magazine* 59:256–68.

Ubaghs, G., 1967. "General characters of Echinodermata," in *Treatise on Invertebrate Paleontology,* Part S., R. C. Moore (ed.). University of Kansas Press, Lawrence, Kan., Echinodermata 1:S3–S60.

Vernon, M. D. (ed.), 1966. *Experiments in Visual Perception*. Penguin Books, Baltimore.

Vrba, E. S., 1980. "Evolution, species and fossils: How does life evolve?" *S. Afr. J. Sci.* 76:61–84.

Weller, J. M., 1961. "The species problem." *Journal of Paleontology* 35:1181–92.

———, 1969. *The Course of Evolution*. McGraw-Hill, New York.

White, M. J. D., 1954. *Animal Cytology and Evolution,* 2nd ed. Cambridge University Press, Cambridge, England.

Woods, H., 1893. *Elementary Paleontology: Invertebrate*. Cambridge University Press. Cambridge, England. 8th ed., 1946.

Wright, S., 1932. "The roles of mutation, inbreeding, crossbreeding, and selection in evolution." *Proceedings of the Sixth International Congress of Genetics* 1:356–66.

———, 1967. "Comments on the preliminary working papers of Eden and Waddington," in *Mathematical Challenges to the Neo-Darwinian Interpretation of Evolution,* P. S. Moorehead and M. M. Kaplan (eds.). *The Wistar Institute Symposia,* Monograph No. 5. The Wistar Institute Press, Philadelphia.

Zittel, K. A. von, 1895. *Grundzüge der Paläeontologie*. R. Oldenbourg Druck und Verlag, Munich.

———, 1896–1902. *Text-book of Paleontology;* trans. and ed. by C. R. Eastman. Macmillan, New York. 2 volumes, in 3 parts. 2nd ed., 1913–1925.